HANDBOOK OF LAND TREATMENT SYSTEMS FOR INDUSTRIAL AND MUNICIPAL WASTES

Handbook of Land Treatment Systems for Industrial and Municipal Wastes

by

Sherwood C. Reed
U.S. Army Corps of Engineers
Cold Regions Research and Engineering Laboratory
Hanover, New Hampshire

Ronald W. Crites
George S. Nolte and Associates
Sacramento, California

np

NOYES PUBLICATIONS
Park Ridge, New Jersey, USA

Library of Congress Catalog Card Number: 84-6025
ISBN 0-8155-0991-X
Printed in the United States

Published in the United States of America by
Noyes Publications
Mill Road, Park Ridge, New Jersey 07656

10 9 8 7 6 5 4 3 2 1

Library of Congress Cataloging in Publication Data

Reed, Sherwood C.
Handbook of land treatment systems for industrial and municipal wastes.

Bibliography: p.
Includes index.
1. Land treatment of wastewater--Handbooks, manuals, etc. 2. Factory and trade waste--Handbooks, manuals, etc. I. Crites, Ronald W. II. Title.
TD760.R45 1984 628.3'6 84-6025
ISBN 0-8155-0991-X

We would like to dedicate this book to the memory of P.H. McGauhey and B.L. Seabrook, and to Richard Thomas and Thomas Hinesly; all of whom were trying to tell us about land treatment before many were ready to listen.

Authors

Sherwood C. Reed is senior environmental engineer for the U.S. Army Cold Regions Research and Engineering Laboratory in Hanover, New Hampshire. He developed and managed a major research program for the Corps of Engineers and, in 1971 prepared a landmark state-of-the-art review on land applications of wastes. He is also special advisor to U.S. E.P.A. headquarters on land treatment, sludge management, innovative and alternative technology. Widely published in this field, he directed the preparation of E.P.A. Design Manuals, and has written numerous articles, and given many lectures and presentations on these topics. He is also a recipient of eight citations from the U.S. Army for achievements related to waste management.

Ronald W. Crites is engineering manager at George S. Nolte and Associates, Sacramento, California. He has prepared studies and designs for numerous municipal and industrial land treatment systems throughout the country. He has also written many professional papers on land treatment, as well as authoring the E.P.A. Process Design Manuals on the subject. In 1982, he won the Wesley W. Horner award from the American Society for Civil Engineers for the best professional paper.

Acknowledgement

We gratefully acknowledge the editorial assistance of Ms. Pamela Crites and Ms. Catherine Girard during preparation of the manuscript; Mr. Edward Perkins for the graphics, and Jon Dyer, George Narita and especially our wives, Susanna and Pam, for their support and encouragement.

Table of Contents

1

Introduction to Land Application As a Treatment Concept

PURPOSE OF THE BOOK

It is the purpose of this book to present the criteria and procedures needed for the complete design of land treatment systems for municipal and industrial wastes. Prior to 1970 land treatment was seldom used by the engineering profession to solve waste treatment problems. In the decade 1970-80 the concept became reestablished in the spectrum of reliable technologies available for wastewater treatment. A considerable research effort has not only established the treatment capability and reliability of land treatment but has also shown that the capital and operating costs and energy usage can be significantly less than for the more familiar mechanical systems. Many of the standard engineering design textbooks still mention land treatment only in passing, if at all. This book is intended to fill that gap.

HISTORICAL BACKGROUND

Disposal of wastes to the land has been an accepted and recognized cultural practice since time began. Stabilization and assimilation of body wastes in the soil are complete, and problems do not occur with low-density, migratory populations of people or animals. The higher density conditions that can cause problems have been documented since biblical times (1), and these problems require a technology for management rather than random disposal. Application of wastes to the land emerged as a technology with the centralization of people in towns and cities. The earliest system documented in the literature was Bunzlau, Germany (2) where a sewage irrigation system was in operation for over 300 years, commencing in 1531. A system in the vicinity of Edinburgh, Scotland began operation around 1650 (2). The value of the wastewater as a fertilizer for vegetable and other crop production was well recognized.

LAND APPLICATION IN NORTH AMERICA

By the mid 19th century land application of wastes was considered to be the safest and most reliable method for waste disposal by the technical experts and regulatory officials of the time. The connection between contaminated water and disease was recognized, although the causative agents were not identified, so waste discharge to water supplies was avoided. The first comprehensive reviews of wastewater disposal in the U.S. were by George Rafter of the U.S. Geological Survey. In a series of reports (3)(4)(5) from 1894 to 1899 he reviewed the status of wastewater treatment in the U.S. and in Europe. Most of the 143 sewage treatment facilities in the U.S. and Canada as of 1899 (3) were land treatment systems (See Table 1-1). Rafter drew the following conclusions (direct quotations):

- The most efficient purification of sewage can be attained by its application to land.
- On properly managed sewage farms the utilization of sewage is not prejudicial to health.
- Sewage may be purified by broad irrigation at all seasons of the year at any place where the mean annual temperature

of the coldest month is not lower than about 20° to 25° F.

- From the experience gained abroad it is clear that we may successfully cultivate almost any of the ordinary agricultural productions of the United States on sewage farms, due regard being had in every case to the special conditions required for each particular crop.

- Sewage utilization should go hand in hand with purification. When operated with reference to all the necessary conditions, a proper degree of purification may be attained as well as satisfactory utilization.

- The proper method of utilizing sewage is, for purposes of irrigation, be means which do not differ, except in matters of detail, from those of ordinary irrigation as practiced abroad for centuries.

Table 1.1: Some Early Land Treatment Systems in the U.S.

Location	Date Started	Area (acres)
Boulder, CO	1890	–
Calumet City, MI	1888*	12
Woodland, CA	1889*	240
Fresno, CA	1891*	4,000
San Antonio, TX	1895	4,000
Vineland, NJ	1901*	14
Lubbock, TX	1915*	–
South Framingham, MA	1889	–

*System still in use.

Current Status

The use of land treatment began to decline soon after Rafter published these conclusions, and by the 1960's the concepts were almost forgotten. By the time discussion again began in the early 70's many of his conclusions were the subject of bitter debate and controversy. Jewell and Seabrook (2) traced the history of the development and the long, but temporary, decline. Among the factors identified for the decline were: pressures for alternative land uses, overloading due to incomplete technical understanding, and the development of the germ theory, with the use of chlorine as an effective disinfectant which made it "safe" to discharge partly treated sewage to waterways.

The focus shifted almost completely to "modern methods of sewage treatment" by the early 1920's, and design criteria for trickling filters, activated sludge, etc. were all available. A considerable effort has been expended in the past 50 years to improve the efficiency and understanding of these "modern methods" but the basic design criteria are still about the same as they were in the beginning. By the late 60's it was recognized that there was more to pollution than BOD and suspended solids and it was decided that a strong federal role and funds would be needed to clean up the nation's waterways. Federal legislation, commencing with the Clean Water Act of 1972 (PL 92-500) proposed a "zero discharge" concept and encouraged a reuse and recovery philosophy. Land application of wastewater is the only way to achieve all of those goals and the concept was reborn. It was not accepted at the time as a technology by much of the engineering profession and regulatory community. As a result, a very significant research and development effort was undertaken to reconfirm the conclusions that were so obvious to Rafter and to develop criteria for reliable and cost effective design, construction and operation. Table 1-2 demonstrates the recent increase in the number of land treatment systems.

Table 1.2: U.S. Municipalities Using Land Treatment

Year	Number of Systems
1940	340
1945	422
1957	461
1962	401
1968	512
1972	571
1981	1,180

In Rafter's time sewage treatment systems were typically found only at the larger, more sophisticated metropolitan centers that could not discharge to an ocean, and land treatment was the preferred method. Except in special circumstances it is unlikely that land treatment would be the preferred method of treatment for the very large metropolitan centers that exist today. The costs and the jurisdictional problems in developing a single very large system would be difficult to overcome. However, there are no technical constraints on the size of a system and, as will be shown in the remaining chapters of this book, land treatment can be a viable and cost effective choice for many industries, for small towns and moderately large cities and for portions of large metropolitan areas.

The design approach for land treatment systems is essentially empirical; that is, observation of successful performance followed by derivation of criteria and mathematical expressions that describe overall performance. Criteria for all of the wastewater treatment processes in use today were developed this way. Use of the criteria in this book should produce reliable, cost-effective, and conservative designs for municipal and industrial wastes.

REFERENCES

1. Deuteronomy, 23:12-14.
2. Jewell, W.J. and Seabrook, B.L., A History of Land Application as a Treatment, USEPA 430/9-79-012, USEPA, OWPO, Wash., D.C., (1979).
3. Rafter, G.W. and Baher, M.N., Sewage Disposal in the United States, 598 pg., (1894).
4. Rafter, G.W., Sewage Irrigation. Water Supply and Irrigation Papers, USGS Rpt. 3, USGPO, Washington, D.C., (1897).
5. Rafter, G. W., Sewage Irrigation. Part II, Water Supply and Irrigation Papers, USGS Rpt. 22, USGPO, (1899).

2

Basic Technology and Design Approach

CONCEPTS

Land treatment is defined as the controlled application of wastes onto the land surface to achieve a specified degree of treatment through natural physical, chemical, and biological processes within the plant-soil-water matrix. The basic wastewater concepts include: Slow Rate (SR), Rapid Infiltration (RI) and Overland Flow (OF). The titles were adopted to reflect the rate of water movement and the flow path within the process. In some earlier texts the term "irrigation" was often used to describe the slow rate process. The present term was chosen to focus attention on wastewater treatment rather than on irrigation of crops. In addition to these basic wastewater processes there are criteria presented in later chapters for combination systems, wetlands and other innovative technologies, large scale septic tank/soil systems, and land application of sludge.

Site Characteristics

The desirable site characteristics for the three wastewater processes are given in Table 2-1. These are not limits to be adhered to rigorously, but rather typical ranges based on successful experience.

Table 2.1: Site Characteristics for Land Treatment Processes

	SR	RI	OF
Grade	20% cultivated 40% uncultivated	Not critical	2-8% Final slopes
Soil permeability	Moderate	Rapid	Slow
Groundwater depth	2-10 ft	3 ft during application 5-10 ft during drying	Not critical
Climate	Storage often needed during cold weather	Not critical	Same as SR

Design Features

Typical design features for the three land treatment processes are compared in Table 2-2. The range of values given represent successful experiences in a variety of locations in the United States. Chapters 10, 11, and 12 contain the procedures for developing site specific design criteria for both planning and final design purposes.

Table 2.2: Typical Design Features for Land Treatment Processes

	SR	RI	OF
Application method	Sprinkler or surface	Usually surface	Sprinkler or surface
Annual loading, ft	2-20	20-400	10-70
Treatment areas for 1 mgd, acres	60-700	7-60	15-110
Weekly application, inch	0.5-4	4-96	2.5-16
Minimum preapplication treatment	Primary	Primary	Grit removal and comminution
Need for vegetation	Required	Sometimes used to stabilize soil	Required

Performance

The expected quality of the treated water from the basic land treatment processes is shown in Table 2-3 for the most common wastewater parameters. The fate of metals, trace elements and salts and the more complex organic compounds are discussed in Chapter 3. These average values result from the treatment that will occur within the immediate plant-soil matrix with no credit for mixing, dispersion or dilution with the groundwater or further travel in the subsoil. Phosphorus, for example, can be reduced at least another order of magnitude for RI systems by additional travel through the soil.

Table 2.3: Expected Water Quality from Land Treatment Processes (mg/ℓ unless otherwise noted)

	SR	RI	OF
BOD	<2	5	10
SS	<1	2	10
NH_3-NH_4 (as N)	<0.5	0.5	<4
Total N (as N)	3	10	5
Fecal Coli #/100 ml	0	10	200+
Total P	<0.1	1	4

SLOW RATE PROCESS

Slow rate (SR) land treatment is the controlled application of wastewater to a vegetated land surface at a rate typically measured in terms of a few inches of liquid per week. The design flow path depends on infiltration, percolation, and usually lateral flow within the boundaries of the treatment site. Treatment occurs at the surface and as the wastewater flows through the plant-soil matrix. A portion of the flow may reach the groundwater, some is used by the vegetation, but offsite runoff of any of the applied wastewater is avoided by the system design. Schematic views of the typical hydraulic pathways for SR treatment are shown in Figure 2-1. Wastewater application can be via ridge and furrow or border strip flooding or via sprinklers using fixed risers or moving systems. The selection of the application method is dependent on site conditions and process objectives and is discussed in detail in Chapters 9 and 10. The surface vegetation is an essential component in the treatment process (Chapter 5); site selection criteria, design details, and management criteria are given in Chapters 6, 10, and 14 respectively.

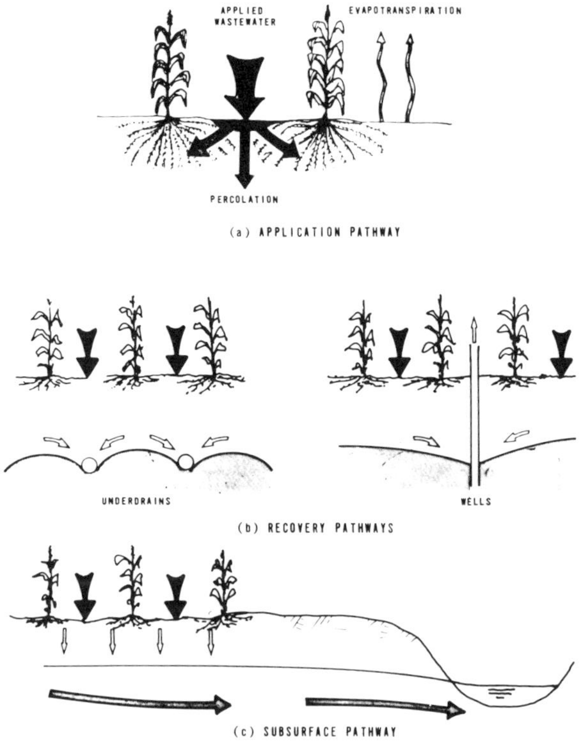

Figure 2-1 Hydraulic pathways for Slow Rate (SR) treatment, After (14).

Slow rate land treatment can be operated to achieve a number of objectives including:

1. Treatment of applied wastewater.
2. Economic return from use of water and nutrients to produce marketable agricultural crops.
3. Exchange of wastewater for potable water for irrigation purposes in arid climates to achieve overall water conservation.
4. Development and preservation of open space and greenbelts.

These goals are not mutually exclusive but it is unlikely that all can be brought to an optimum level within the same system. In general, maximum cost effectiveness for both municipal and industrial systems will be achieved by applying the maximum possible amount of wastewater to the smallest possible land area. That will in turn limit the choice of suitable vegetation and possibly the market value of the harvested crop. In the more humid parts of the United States optimization of treatment is usually the major objective. Optimization of agricultural potential or water conservation goals are generally more important in the arid western portions of the United States.

Optimization of a system for wastewater treatment usually results in the selection of perennial grasses because a longer application season, higher hydraulic loadings, and greater nitrogen uptake are possible compared to other agricultural crops. Annual planting and cultivation can also be avoided with perennial grasses. However, corn and other crops with higher market values are also grown on systems where treatment is a major objective. Muskegon, MI (1) is a noted example with over 5,000 acres of corn under cultivation.

Forested systems also offer the advantage of a longer application season and higher hydraulic loadings than typical agricultural crops but may be less efficient than perennial grasses for nitrogen removal depending on the type of tree, its stage of growth and the general site conditions. Research at Pennsylvania State University (2) established the basic criteria for full scale forested systems. Similar work in Georgia, Michigan and Washington further refined the criteria for regional and species differences (3). A large scale slow rate forested system in Clayton County, Georgia, designed for 19.6 mgd, has been in operation since 1981 (4).

RAPID INFILTRATION PROCESS

Rapid infiltration (RI) land treatment is the controlled application of wastewater to earthen basins in permeable soils at a rate typically measured in terms of feet of liquid per week. As shown on Table 2-2 the hydraulic loading rates are usually at least an order of magnitude higher than SR systems. Any surface vegetation that is present has a marginal role for treatment performance due to the high hydraulic loading. However, vegetation is sometimes critical for the stabilization of surface soils and the maintenance of acceptable infiltration rates. Treatment is accomplished by biological, chemical and physical interactions in the soil matrix with the near surface layer being the most active zone.

The design flow path requires infiltration, percolation and typically lateral flow away from the application site. A cyclic application as described in Chapter 12 is the typical mode of operation with a flooding period followed by days or weeks of drying. This allows aerobic restoration of the infiltration surface and drainage of the applied percolate. The geohydrological aspects of the RI site are more critical than for the other processes and a proper definition of subsurface conditions and the local groundwater system is essential for design. Schematic views of the typical hydraulic pathways for RI treatment are shown in Figure 2-2.

The purpose of a rapid infiltration system is wastewater treatment and the system design and operating criteria are developed to achieve that goal. However, there are alternatives with respect to the utilization or final disposal of the treated water:

1. Groundwater recharge.
2. Recovery of treated water for subsequent reuse or discharge (Figure 2-2b).
3. Recharge of surface streams (Figure 2-2c).
4. Seasonal storage of treated water beneath the site with recovery and agricultural reuse during the growing season.

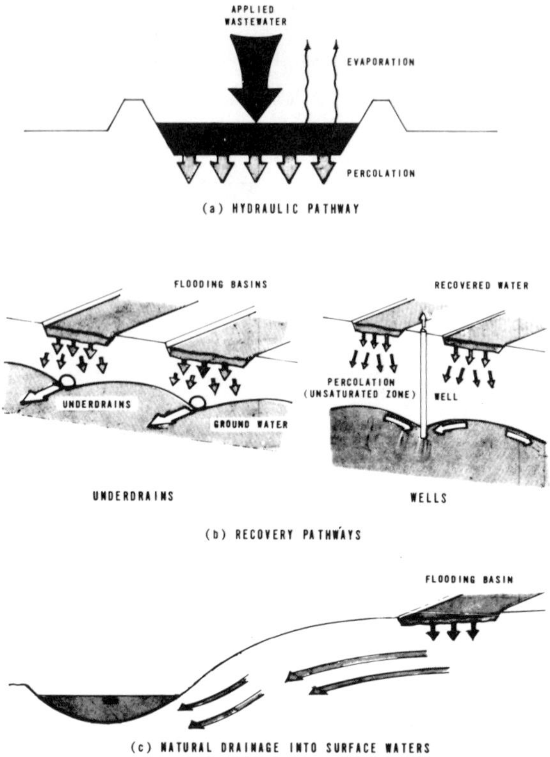

Figure 2-2 Hydraulic pathways for Rapid Infiltration (RI) treatment, After (14).

The recovery and reuse of the treated wastewater is particularly attractive in the more arid parts of the U.S. and studies in Arizona, California, and Israel (5)(6)(7) have demonstrated that the quality of the recovered water is suitable for unrestricted irrigation on any kind of crop. Groundwater recharge may also be desirable but particular attention is required for the nitrogen component in the applied wastewater. The high hydraulic loading rates commonly employed make it unlikely that drinking water levels for nitrate nitrogen (10 mg/L as N) can be attained immediately beneath the application zone with typical municipal wastewaters. If recharge of a drinking water aquifer is planned there must then be sufficient mixing and dispersion with the insitu groundwater prior to the down gradient extraction points. In the more humid parts of the U.S. neither recovery and reuse or groundwater recharge are typically necessary. An exception could be groundwater recharge for prevention of salinity intrusion, as practiced on Long Island. In most cases the RI site is often selected so that subsurface flow will emerge in an adjacent surface water as shown on Figure 2-2c. The quality of this sub flow entering the surface water will generally exceed that which could be produced by a mechanical AWT plant.

OVERLAND FLOW PROCESS

Overland flow (OF) is the controlled application of wastewater to relatively impermeable soils on grass covered gentle slopes. The hydraulic loading is typically several inches of liquid per week and is usually higher than for most SR systems. Since system costs tend to be directly related to hydraulic loading, OF systems are usually more cost effective than SR systems, for the same water quality requirements. Vegetation, consisting of perennial grasses, is an essential component in the system, both for its physical contribution to slope stability and erosion protection and for its contribution to nutrient and other contaminant removal.

The design flow path is essentially sheet flow down the carefully prepared vegetated surface with runoff collection in ditches or drains at the toe of the slope. Treatment occurs as the applied wastewater interacts with the soil, vegetation, and biological surface growths. Physical, chemical, and biological reactions are all involved. A small portion of the water may be lost to deep percolation and a larger fraction is lost by evapotranspiration, but the major part of the applied wastewater is collected in the ditches and is discharged, typically to an adjacent surface water. A typical site and the hydraulic pathway involved are illustrated in Figure 2-3.

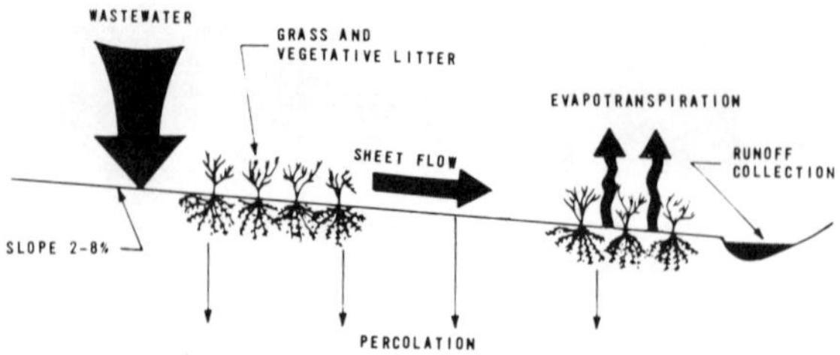

(a) HYDRAULIC PATHWAY

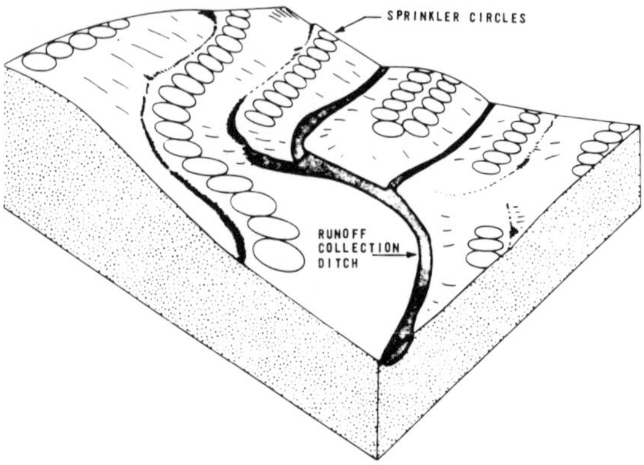

(b) PICTORIAL VIEW OF SPRINKLER APPLICATION

Figure 2-3 Hydraulic pathways for Overland Flow (OF) treatment, After (14).

The SR and RI processes may include drainage/recovery and discharge, but the OF concept almost always incorporates a surface discharge, and permits for discharge are required.

The purpose of overland flow is cost effective wastewater treatment. The harvest and sale of the cover crop may provide some secondary benefit and help to offset operational costs, but the primary objective is treatment of the wastewater:

1. Overland flow may be the basic treatment component in a system designed to achieve secondary effluent quality with minimal preapplication treatment.
2. Overland flow may be used as a polishing or AWT step for high levels of BOD, SS, and nitrogen removal.

A rational design approach is presented in Chapter 11 that will permit reliable system design to attain either of these objectives. In essence, the OF slope is similar in concept to any other attached growth, fixed film biological process for the removal of BOD, SS, and nitrogen. The addition of grass to the system as a vegetative component further enhances nitrogen removal. The largest municipal overland flow system is in Davis, CA., designed for 5 mgd flow (8). Oxidation pond effluent is applied using gated pipe. Field tests at the same site with primary effluent demonstrated at least comparable and often better and more consistent treatment performance.

LIMITING DESIGN PARAMETER

The design of all land treatment processes, and that of many other engineering works, is based on the Limiting Design Parameter (LDP) concept. It is necessary with this approach to identify the critical factor or parameter that limits or controls the design. Successful performance is then insured by designing for that parameter so that all other parameters are also under control.

The LDP for mechanical treatment plants is typically a wastewater constituent such as BOD, SS, N or P. Adjustments for the volume of flow are made by increasing the size or the number of the units but process design is still based on the wastewater constituent. However, the volume of flow or hydraulic loading can be an independent and controlling variable for some land treatment systems. For example, a grass covered hay field might be able to receive 3 in/wk of wastewater and still satisfy nitrogen requirements in the percolate. However, if the soil were a slowly permeable clay the application rate might be limited to 1 in/wk or less due to the hydraulic constraints. In this case the hydraulic capacity would be the LDP for design since the nitrogen requirement should automatically be satisfied at the lower hydraulic loading.

The hydraulic capacity also tends to be the LDP for design of RI systems since nitrogen is often excluded as a controlling parameter either because the percolate is recovered or the groundwater impacts are acceptable.

Detailed discussion on the interactions in land treatment systems with the major wastewater constituents can be found in Chapter 3. Experience has shown that land treatment systems which depend on significant infiltration (SR,RI) have either nitrogen or hydraulic capacity as the LDP when typical municipal wastewaters are applied (10). Overland flow may have either BOD, SS, or N as the LDP depending on the particular site specific discharge requirements. In all three cases, all of the other wastewater constituents can be adequately managed if the system is designed on the basis of the LDP. Overland flow is similar to the mechanical systems in two respects, first the LDP controls process design and the number of units is then adjusted to match predicted flow. Second, pathogen removal is not considered a limiting factor in process design. Pathogen removal will occur in overland flow, but final disinfection may be required to satisfy stringent discharge standards as with mechanical systems. The LDP for the basic land treatment concepts are summarized in Table 2-4.

Table 2.4: Limiting Design Parameters (LDP) for Typical Municipal Effluents

Land Treatment Process	LDP
Slow Rate (SR)	Hydraulic capacity of soil or nitrate contamination of aquifer
Rapid Infiltration (RI)	Usually hydraulic capacity of soil
Overland Flow (OF)	Whichever parameter (BOD, or SS, or N) in discharge permit that requires the longest travel distance/detention time for removal

Determination of the LDP for industrial wastewater can be much more complex since it may be unique for a particular production facility and the waste streams involved. In general wastewaters from food processing industries tend to be similar to municipal effluents in that the amounts of metals, trace organics, and toxic substances tend to be relatively low. The organic content and SS may be high but as discussed in Chapter 3, as long as the organics are easily degradable even very high BOD loadings will not become the constraint on design. Some systems with minimal pretreatment may have suspended solids (SS) as the LDP if these solids are slowly degradable and would accumulate and clog the soil surface.

General "rules of thumb" for determining the LDP for most industries are not possible due to the unique characteristics for each industry, compounded by the process variabilities from site to site within the same basic industry. Determination of the LDP starts with identification of the groundwater and surface water parameters of concern to the regulatory agencies. In the former case if drinking water aquifers are involved then drinking water standards will apply. In addition, the long term accumulation of materials on the site must be considered in the context of the future intended use. Chapter 3 and other sources (11)(12)(13)(14) can be consulted to determine the LDP for a particular industry.

REFERENCES

1. Demirjiian, Y.A., et. al. Muskegon County Wastewater Management System Progress Report 1968-1975, EPA 905/2-80-004 USEPA Region V, (Feb. 1980).
2. Kardos, L.T., et. al., Renovation of Secondary Effluent for Reuse as a Water Resource, EPA 660/2-74-016 (Feb. 1974).
3. McKim, H.L., et. al., Wastewater Applications in Forest Ecosystems, USA CRREL Report 82-19, (August 1982).
4. Crites, R.W. Slow Rate Land Treatment: A Recycle Technology. USEPA, OWPO (1981).
5. Bouwer, H. et. al. Rapid-Infiltration Research at Flushing Meadows Project, Arizona, Journal WPCF, 52(10), 2457-2470, (Oct. 1980).

6. Olson, J.V., Crites, R.W., Levine, P.E. Groundwater Quality at a Rapid Infiltration Site. Journal EED, ASCE, 106(5): 885-299. (1980).

7. Idelovitch, E. Unrestricted Irrigation with Municipal Wastewater Proceedings, National Conference on Environmental Engineering, ASCE, Atlanta, Georgia (July 8-10, 1981).

8. Smith, R.G. and Schroeder, E. D. Physical Design of Overland Flow Systems, Journal WPCF, 55(3), 255-260, (March 1983).

9. Black, S.A., Wile, I., Miller, G. Sewage Effluent Treatment in an Artificial Marshland Presented at the Annual Conference of the WPCF, Detroit, MI, (Oct. 1981).

10. Reed, S.C., Design, Operation and Maintenance of Land Application Systems, Proceedings-Workshop on Low Cost Wastewater Treatment, Clemson University (April 1983).

11. Overcash, M.E., Pal, D., Design of Land Treatment Systems for Industrial Water-Theory and Practice, Ann Arbor Science, Ann Arbor, MI. (1979).

12. Middlebrooks, E.J., Industrial Pollution Control-Vol. 1: Agro Industries. John Wiley & Sons, NY, NY (1979).

13. Parr, J.F., Marsh, P.B., Kla, J.M., Land Treatment of Hazardous Wastes. Noyes Publications, NJ (1983).

14. USEPA, Process Design Manual for Land Application of Municipal Wastewater, US Environmental Protection Agency. EPA 625/1-81-013 (Oct. 1981).

3

Wastewater Parameters and System Interactions

The design approach for any land treatment system is based on the Limiting Design Parameter (LDP) as introduced in Chapter 2. Land treatment differs from mechanical technologies in that removal of metals and some other materials is very effective but these substances can then remain within the soil matrix and their accumulation may limit the useful life of the site and/or its future use for agricultural purposes.

An understanding of the basic interactions between the wastewater parameters and the soil treatment system is essential for the determination of the LDP for a particular design. The parameters of concern can be grouped in nine major categories:

- BOD
- Suspended Solids
- Pathogenic Organisms
- Oils and Grease
- Metals
- Nitrogen
- Phosphorus
- Inorganic Trace Elements and Salts
- Trace Organics

BOD

All land treatment concepts are very efficient at removal of biodegradable organics. Removal mechanisms include filtration, adsorption and biological oxidation. Most of the responses in SR and RI systems occur at the surface or in the near surface soil matrix where microbial activity is most intense. Part of the reason for the intermittent or cyclic wastewater applications on these systems is to allow the restoration of aerobic conditions for biological oxidation. Essentially all of the responses in overland flow occur at the surface. Settling of particulate matter is a contributing factor as the applied wastewater flows in a thin film down the slope. The biological growths and slimes that develop on the slope are primarily responsible for ultimate removal. These growths are similar to those found in other fixed film processes (ie: trickling filters, etc.) so the presence of aerobic and anaerobic zones or microsites within the slime layer is to be expected. In a properly managed system the aerobic zones dominate, but there are still anaerobic zones that contribute to the breakdown of the more refractory organics and contribute to nitrogen removal via denitrification. Typical removals of BOD at land treatment systems receiving municipal wastewater are summarized on Table 3-1. Since the final treatment mechanism is biological, the systems have a continually renewable capacity for BOD removal.

Table 3.1: BOD Removal at Typical Land Treatment Systems (1)-(8)

Process/Location	Wastewater Loading (ft/yr)	..... BOD (mg/ℓ) ..... Applied	Treated Water	Sampling Depth (ft)
SR				
Hanover, NH	4-25	40-92	0.9-1.7	5
San Angelo, TX	10	89	0.7	7
RI				
Lake George, NY	140	38	1.2	10
Phoenix, AZ	360	15	0-1.0	30
Hollister, CA	50	220	8.0	25
OF				
Hanover, NH	25	72	9	–
Easley, SC	27	200	23	–
Davis, CA	41	112	10	–

Organic Loading

A comparison of the values in Table 3-1 indicates that land treatment systems have a very high capacity for treatment of degradable organics. The rapid infiltration systems have a percolate quality close to that of the SR systems but the organic loading is typically an order of magnitude higher than at the SR sites. Additional data from industrial operations would indicate that the RI operations on Table 3-1 are not being stressed by the BOD loadings cited. It has been demonstrated, for example, that a site with acclimated soil bacteria could accept a 10,000 lb/acre·d COD loading satisfactorily (9).

A study at five SR sites applying potato processing wastewater in Idaho demonstrated COD loading rates ranging from 40 to 280 lb/acre·d with removals of up to 98% after 5 ft of percolation (10). Pilot scale OF with high strength snack food processing wastewaters was successful at BOD loading rates ranging from 50 to 100 lb/acre·d (11). Pilot RI studies in Montana with partially treated Kraft process paper mill wastes with BOD loadings up to 600 mg/L at hydraulic loadings of about 0.2 ft/d were also successful (12).

Since some of the industrial systems discussed above successfully operate with applied BOD concentrations of 1000 mg/L or more it is obvious that land treatment of municipal wastewater (at 200-300 mg/L BOD) should be no problem. Therefore, neither BOD or COD will be the limiting factor for design of municipal land treatment systems. Typical organic loading rates in use are summarized in Table 3-2.

Table 3.2: Typical Organic Loading Rates for Land Treatment Systems (13, 14)

Process	Organic Loading (lb BOD/acre·d)
SR	45-450
RI	130-890
OF	35-100

The U.S. EPA recommends an upper limit of 600 lb BOD_5/acre·d for RI systems (14). In the typical municipal system some other parameter (Nitrogen, etc.) is likely to be the LDP and as a result, the organic loadings seldom exceed 100 lb/acre·d. For example, the typical BOD loading rate for SR systems receiving municipal secondary effluent is less than 5 lb/acre·d. This is an order of magnitude lower than the lower limit in Table 3-2 so it can be concluded for at least municipal systems that BOD loadings will not limit design or even be an issue of concern.

SUSPENDED SOLIDS

Slow rate and rapid infiltration systems are very effective in removal of suspended solids. Filtration in the soil matrix is the primary removal mechanism. Overland flow systems depend on sedimentation and entrapment in the vegetative litter or on the biological slimes and are typically less efficient than SR or RI. However, OF systems can typically provide better than secondary effluent quality for suspended solids when screened raw sewage or the equivalent of primary effluent is applied. Table 3-3 summarizes suspended solids removal performance at a number of systems receiving municipal wastewaters.

Table 3.3: Suspended Solids Removal at Land Treatment Systems (6, 7, 15)

Location	.... Suspended Solids (mg/ℓ)	
	Applied	Treated Water
Slow Rate		
Hanover, NH	60	<1
Typical value	120	<1
Rapid Infiltration		
Phoenix, AZ	20-100	0
Hollister, CA	274	10
Typical value	120	2
Overland Flow		
Ada, OK (Raw)	160	8
Hanover, NH (Primary)	59	7
Easley, SC (Raw)	186	8
Utica, MS (lagoon)	30	8
Davis, CA (lagoon)		
Summer	121	80
Fall	86	24
Winter	65	13

There has been mixed experience with respect to suspended solids removal at overland flow systems receiving lagoon effluents, primarly due to the variability of the algal species present. This is discussed in detail in Chapter 11.

Municipal Systems

Most of the suspended solids in municipal effluents are degradable organics in concentrations ranging from 30 to about 350 mg/L depending on the degree of treatment provided prior to land application. Since this is a component in the total organic loading discussed in the previous section, it is clear that the suspended solids in typical municipal wastewaters applied at typical rates should not be the limiting factor for process design. Experience at operating systems indicates the best performance with the least possible degree of preapplication treatment. The solids from screened raw sewage or primary treatment are more easily separated and oxidized than the more refractory secondary effluents or algal-laden lagoon effluents.

Industrial Systems

Problems have occurred in OF systems due to unequal deposition of suspended solids on the treatment slope. The most cost effective systems use gated pipe or other low pressure devices for wastewater distribution at the top of the slope, and will result in deposition of most of the suspended matter in the first 10-15 ft of travel. This will not create objectionable anaerobic conditions with typical municipal wastewaters applied on the cyclic schedules discussed in Chapter 11. High pressure sprinklers should be used for application of high strength, high solids industrial wastewaters to insure a more uniform distribution.

The accumulation of the more refractory solids on the soil surface in SR and RI systems has resulted in clogging problems and a reduction in the expected infiltration rates. These solids might be algal cells, as were observed at an RI system in Phoenix, AZ, or non-to-slowly degradable solids from industrial operations. Loading rates for the latter case can still be quite high.

An SR system in Pennsylvania receiving wastewater from a hardboard production facility was successfully operated at a loading of about 550 lb of solids/acre·d (16). The waste stream consisting of hexosans, pentosans, and hemicellulose products has a BOD ranging from 6000 to 18,000 mg/L. The LDP for operation of this site was developed in terms of pounds of solids per acre per day rather than hydraulic loading or some other wastewater constituent. The loading rate was gradually increased to 900 lb/acre·d when toxic effects were noted. A continuous loading rate of 550 lb solids/acre·d was established on a year round basis with the Reed canarygrass covered site capable of accepting shock loads of up to 700 lb solids/acre·d for brief periods during the summer. Commercial fertilizers were added twice annually since the effluent was deficient in all nutrients.

The 550 lb solids/d represented an organic loading of about 500 lb BOD/acre·d. This represents at least 100 tons of organic solids/acre per year. In contrast, a typical municipal wastewater (200 mg/L SS) applied at typical hydraulic loading rates (10 ft/yr) on an SR site would have a solids loading less than 3 tons/acre per year.

PATHOGENIC ORGANISMS

The pathogens of concern in land treatment systems are: bacteria, parasites, and virus (17). The pathways or vectors of concern are to groundwater, internal or external contamination of crops, translocation to grazing animals and off site transmission via aerosols or runoff. Removal of pathogenic organisms in land treatment is accomplished by adsorption, dessication, radiation, filtration, predation and exposure to other adverse conditions.

The SR system is the most effective removing about 5 logs (10^5) of fecal coliform; RI typically removes 2 to 3 logs and OF about 90% of the applied fecal coliforms (18).

Parasites

Parasites may be present in all municipal wastewaters. Ascaris, E histolytica, Helminths and other parasitic types have been recovered from sewages and sludges. Under optimum conditions the eggs of these parasites, particularly Ascaris can survive for many years in soil. Because of their weight, parasite cysts and eggs will settle out in preliminary treatment or storage ponds so most will be found in sludges.

There is no evidence available indicating transmission of parasitic disease from application of wastewater in land treatment systems. Hays (19) described several situations where the use of raw sewage in gardens was the suspected source of disease. Such a practice is outside the limits recommended so should not be a factor in the United States. Transmission of parasites via aerosols should not be a problem due to the weight of the cysts and eggs. There is some evidence of animal infection. Beef measles and tapeworm have been documented (19) but are due to direct contact or dinking of the wastewater by animals. Schistosomiasis, which is a very serious parasitic problem in many parts of the world due to direct contact by humans with polluted waters, is not in the continental U.S. because the alternate host snails are not present. Dean (20) indicates that Ascaris eggs have occasionally been found in the soil at the Paris, France sewage farm which receives essentially raw sewage but that vegetables produced on the farm are approved by the Ministry of Health and sold on the open market in Paris. Vegetables to be eaten raw are not permitted.

Crop Contamination

Concerns with respect to crop contamination focus on surface contamination and then persistance of the pathogens until consumed by man or animals or the internal infection of the plant via the roots. The persistance of polio virus on lettuce and radishes for up to 36 days has been demonstrated (17). About 99% of the detectable virus were gone in the first 5 to 6 days. The general policy in the U.S. is not to grow vegetables to be consumed raw at land treatment systems to eliminate such risks. Shuval described internal contamination of plants with virus via transport from the roots to the leaves. However, these results were obtained with soils innoculated with high concentrations of virus and then the roots were damaged or cut. No contamination was found when roots were undamaged or when soils were not innoculated with the high concentrations.

Criteria for irrigation of pasture with primary effluent in Germany require a period of 14 d after sprinkling before animals can be allowed to graze (21). Experiments by Bell (22) in Canada indicated that fecal coliforms on the surfaces of alfalfa hay, from sprinkling with wastewater, were killed by 10 h of bright sunlight. He also experimented with Reed canarygrass and found 50 h of sunlight required. The longer period required is probably due to the sheath on the grass leaf which is not present on alfalfa. He recommended a 1 wk rest period prior to grazing to insure sufficient sunlight, for Reed Canary, orchard and brome grasses used for forage or hay. Since fecal coliforms have similar survival characteristics to salmonella, he suggests results should be applicable to both organisms.

Runoff Contamination

Runoff from a land treatment site might be a potential pathway for pathogen transport. A proper system design would eliminate runoff from adjacent lands entering the site and runoff of applied wastewater, except for the OF case. Runoff of renovated effluent is a design feature of overland flow. It is controlled runoff, collected in ditches prior to surface discharge or further treatment. Deemer (23) indicates that the quality of rainfall runoff from overland flow sites is equal to or better in quality than the normal renovated wastewater runoff. Runoff is not a factor of concern for rapid infiltration systems. For slow rate systems if proper measures for erosion control are taken then runoff quality, if any occurs, should be no different than expected from normal agricultural practice.

Groundwater Contamination

The risk of groundwater contamination by pathogens involves movement of bacteria or virus to aquifers that are then used for drinking purposes without further treatment. The risk is not an issue for OF systems but has the highest potential for RI due to the high hydraulic loading rates and coarse textures of the receiving soils.

Bacteria removal in the finer textured agricultural soils commonly used for slow rate systems can be quite high. Results from a five year study in Hanover, NH (17) applying both primary and secondary effluents to two different soils indicated essentially complete removal of fecal coliforms within a 5 ft soil profile. The soils involved were a fine textured silt loam and a coarser textured loamy sand, concentrations of fecal coliforms in the applied wastewater ranged from 10^5 for primary effluents to 10^3 for secondary effluents. In similar research in Canada (22), undisinfected effluent was applied (1.8 in./wk) to grass covered loamy sand. Most of the fecal coliforms were retained in the top 3 in. of soil, none penetrated below 27 in. Die-off occurred in two phases: an initial rapid phase within 48 h of application during which 90% of the bacteria died, followed by a slower decline during a two week period when the remaining 10% were eliminated. Removal of virus which is dependent on adsorption reactions should also be quite effective on these finer textured soils.

Most of the concern, and the research work on virus transmission through soils, has focused on RI systems and coarse textured soils. Table 3-4 is a summary of results from several studies.

The RI basins in the Phoenix work consisted of about 30 in. of loamy sand underlain by coarse sand and gravel layers. During the study period indigenous virus were always found in the applied wastewater but none were recovered in the sampling wells as shown in Table 3-4.

Table 3.4: Virus Transmission Through Soil at Land Application Sites (17)

Location	Sampling Depth/Distance (ft)	. . Virus Concentration (pFu/ℓ). . . At Source	At Sample Point
Phoenix, AZ	10-30	8	0
(Jan-Dec 1974)		27	0
		24	0
		2	0
		75	0
		11	0
Gainesville, FL	23	0.14*	0.005
(April-Sept 1974)		0.14	0
		0.14	0
		0.14	0
		0.14	0
		0.14	0
		0.14	0
		0.14	0
Santee, CA (1966)	200	Concentrated Type 3 polio	0

*Average.

At Santee, CA secondary effluent was applied to percolation beds in a shallow stratum of sand and gravel (24). Percolate moved to an interceptor trench approximately 1500 ft from the beds. Enteric virus were isolated from the treatment ponds but none ever found at observation wells at 200 ft and 400 ft from the site.

Lance and others have examined the problem of virus desorption in the laboratory. Using soil columns it can be shown that applications of distilled water or rainwater can cause adsorbed virus to move deeper into the soil profile under certain conditions. In Lance's work viruses were not desorbed if the free water in the column drained prior to application of the distilled water. This suggests that the critical period would be the first day or two after the wastewater infiltrated. Rainfall after that period should not then cause movement of virus. Even if some movement does occur, the soil profile in nature does not necessarily have a finite bottom like a soil column. A desorbed viral particle should have further opportunities for readsorption in the natural case. Lance's work with polio virus in soil columns (calcareous sand) indicates that most virus particles are retained near the soil surface (26). Increasing the hydraulic loading from 2 ft/d to 4 ft/d caused a virus break through (about 1% of the particles added) at the bottom of the 8 ft column. However, 99% was still removed at hydraulic loadings as high as 39 ft/d. Lance suggests that the velocity of water movement through the soil may be the single most important factor affecting the depth of virus penetration (26).

Aerosols

The potential for aerosol transport of pathogens from land treatment has probably been one of the most controversial health issues. The concerns go beyond technical and health aspects and involve aesthetics and emotions. The lay public, and many professionals, tend to misunderstand what aerosols are and confuse them with the droplets that emerge from the sprinkler nozzles. Aerosols are almost colloidal in size ranging from 20 microns in diameter or smaller. It is prudent to design any land treatment system so the larger droplets are contained within the site. The public acceptance of a project will certainly be enhanced if it is understood that neither their persons nor their property will become "wet" from the sprinkler droplets.

Bacterial aerosols are present in all public situations and will tend to increase with the number of people and their proximity. Sporting events, theaters, public transportation, public toilets, etc., are all potential locations for airborne infection.

Table 3.5: Aerosol Bacteria at Various Sources (17)

Source	Downwind Distance (ft)	Total Aerobic Bacteria (#/ft^3)	Total Coliform (#/ft^3)
Activated sludge tank Chicago	30-100	396	0.2
Activated sludge tank Sweden	0	2,832	–
Power plant cooling tower (disinfected), California	0	83	–
Mechanical aerated pond Israel	100	–	8
Sprinklers (disinfected) Ohio	100	14	0.1
Sprinklers (undisinfected) Israel	100	–	3.3
Sprinklers (undisinfected) Arizona	150	23	0.2
Sprinklers (undisinfected) Pleasanton, California	30-100	73	0.2

Note: Aerosol counts are per cubic foot of air.

Data in Table 3-5 summarizes aerosol bacteria concentrations at various sources, all of which involve the use or treatment of wastewaters. The cooling water for the power plant that is cited uses some disinfected effluent as make-up water. It is interesting to note that the aerosol concentration at this cooling tower is roughly the same as measured just outside the sprinkler zone in Pleasanton, CA where undisinfected effluent is used. It does not appear from these data that the bacterial aerosols at or near land treatment sites are worse than other sources. In fact, the opposite seems true, the aerated pond in Israel and the activated sludge systems cited have higher concentrations than the comparable land treatment sites. Adams (28) cites aerosol studies in metropolitan areas and indicates a bacterial concentration of $4/ft^3$ of air during the day time in Louisville, KY, and an annual average of $57/ft^3$ of air (range 30-84) for Odessa, Russia. The bacterial aerosols from land treatment systems listed in Table 3-4 fall within this range.

An epidemiological study at an activated sludge plant in the Chicago area (29) documented bacteria and virus in aerosols at the plant. However, microbial and chemical content of the air, the soil, and the surface waters in the surrounding area were not different than background levels and no significant illness rates due to the activated sludge plant were revealed within a 3 mile radius. A similar effort was undertaken at an activated sludge plant in Oregon with a school playground approximately 30 ft from the aeration tanks. Positive counts for aerosol bacteria were noted in the school yard but no adverse health responses in the children. It can be inferred from these studies that since the concentrations of virus and bacteria in aerosols from land treatment are equal to or less than the concentrations from activated sludge and since there were no adverse health effects in the latter case there then should not be any for land treatment.

Work in Israel (17) also documents the comparability of land treatment versus conventional modes. Table 3-6 summarizes downwind coliform concentrations from sprinklers and a pond with surface aerators.

Table 3.6: Aerosol Coliforms from Sprinkling and Aerated Pond in Israel (17)

	.... Concentration ($\#/ft^3$) at Distance..........				
Source	100 ft	330 ft	490 ft	650 ft	820 ft
Aerated pond*	8	0.1	0.1	0	0
Sprinklers**	3	0.6	0.2	0	0.1

*Temperature 85°F, humidity: 75%, Solar Rad 7-8, wind 20 ft/s Endo medium.
**Temperature 75°F, humidity: 75%, Solar Rad 0-6, wind 15 ft/s Endo medium.

The aerosol measurements at the Pleasanton, CA land treatment system (29) demonstrated that salmonella and viruses survived longer than did the traditional coliform indicators. However, the downwind concentrations of virus were very low, 0.0004 pfu/ft^3. The source was undisinfected effluent from high pressure impact sprinklers and the sampling point was about 160 ft from the sprinklers. The concentration cited is equal to one virus particle in every 250 ft^3 of air. Assuming a normal breathing intake of approximately 0.7 ft^3/min it would take 59 h of continuous exposure for the operator to inhale that much air. It is estimated that the operators might spend about an hour per work day that close to the sprinklers which is equivalent to the time spent servicing the aeration tanks at an activated sludge plant. At that rate the operator at Pleasanton would be exposed to less than 4 virus per year and the risk to the adjacent population would seem to be non-existent.

The current EPA guidance recommends a fecal coliform count of 1000/100mL for recreational type application. These two criteria are based on pre-existing standards for general irrigation water and for bathing waters and body contact sports. With respect to the aerosol risk of sprinkling such waters, Shuval (31) has reported that when the coliform concentration at the nozzle was below 1000/100 mL there were none detected at downwind sampling stations.

Procedures have been developed for estimating the downwind concentration of microorganisms in aerosols from sprinkler application of wastewater (32). The equation takes the form:

$$C_d = C_n D_d e^{(xa_d)} + B \qquad (3\text{-}1)$$

where: C_d = concentration at distance d, #/ft^3

C_n = microorganisms released at source, #/s

D_d = atmospheric diffusion factor

x = decay, or die-off rate, s^{-1}

$$a_d = \frac{\text{downwind distance (d)}}{\text{wind velocity}}, \quad \frac{(ft)}{ft/s}$$

B = background concentration in upwind air.

The microorganisms released at the source (C_n) is a function of the density in the wastewater, the wastewater flow rate, the aerosolization efficiency and a survival factor:

$$C_n = W\ F\ E\ I \tag{3-2}$$

where

C_n = organism release at source (#/s)

W = density in the wastewater (#/L)

F = flow rate (L/s) or gpm x 0.06308

E = aerosolization efficiency

I = survival factor

The survival factor ranges from about 0.27 for fecal coliforms to 80 for virus (32). Research at a number of sites indicate that with moderate to high pressure sprinklers about 0.33% (E = .0033) of the wastewater is converted to aerosol droplets (32). The decay rate (x) is about -0.023 for fecal coliform and is assumed to be zero for virus. The atmospheric dispersion factor (D_d) depends on a number of related meteorological conditions. Typical values for a range of expected conditions are given in Table 3-7, reference (32) should be consulted if a more exact determination is needed.

Table 3.7: Atmospheric Dispersion Factor for Aerosols at a Distance of 300 feet from the Source (Adapted from Reference 32)

Atmospheric Conditions	D_d (s/ft^3)
Windspeed <4 mph Strong sunlight	5×10^{-6}
Windspeed <4 mph Cloudy daylight	11×10^{-6}
Windspeed 4-10 mph Strong sunlight	4×10^{-6}
Windspeed 4-10 mph Cloudy daylight	9×10^{-6}
Windspeed >10 mph Strong sunlight	8×10^{-6}
Windspeed >10 mph Cloudy daylight	17×10^{-6}
Windspeed >7 mph Night	17×10^{-6}

Example 3-1: Aerosols

Condition: Sprinkler (100 ft diameter circle) discharging at 1100 gpm: Aerosolization Efficiency = 0.0033, Survival factor (fecal coliforms) = 0.27, Decay rate (fecal coliforms) = -0.023.

Fecal coliforms in wastewater 10^5/L; sprinklers operating during daylight hours with strong solar radiation, and a wind speed of 5 mph. Background coliforms = 0. Find the fecal coliform concentration at 300 ft downwind of sprinkler.

Solution.

$$C_n = W\ F\ E\ I$$
$$= (10^5)(1100)(0.0631)(0.0033)(0.27)$$
$$= 6184/\text{s, released}$$
$$D_d = 4 \times 10^{-6} \text{ (from Table 3-7)}$$
$$a_d = \frac{300 \text{ ft}}{7.33 \text{ ft/s}} = 40.9 \text{ s}^{-1}$$
$$X = -0.023$$
$$B = 0$$
$$C_d = QD_d e^{(X \cdot a_d)}$$
$$= (6184)(4 \times 10^{-6})\ e^{-0.023(40.9)}$$

$= (6184)(4x10^{-6})(0.39)$

$=$ 0.002 fecal coliforms per cubic foot of air at a distance of 300 ft downwind of sprinkler (200 ft from wetted circle).

OILS AND GREASE

Oils and grease should not be a factor for land treatment of typical municipal wastewaters unless there is an accidental spill somewhere in the municipal collection system. However, there is no need to design the basic system for such an emergency since containment and clean-up procedures can be used if needed.

Oils and grease are more likely to be a routine component in industrial wastewaters. The three major sources are petroleum and animal and vegetable oils. Food processing, rendering, soap manufacturing, and margarine and wax production are all sources of animal or vegetable oils. Wastewaters from seafood processing can have up to 12,000 mg/L free and/or emulsified oil and grease (33). The intentional discharge of petroleum products to sewers is not expected but leaky devices and the wash down of equipment and facilities can result in a significant contribution. The oil concentration ranged from 23 to 130 mg/L in wastewaters from 12 refineries (34).

Most of the research and study that has been directed at the interactions between oil and the soil-plant ecosystem has been directed at petroleum products. The major purpose of many of those efforts has been to gain a better understanding of the effects of an oil spill and to develop criteria for restoration and/or mitigation. Two pathways for oil removal in a natural system have been demonstrated. The volatile portion is lost to the atmosphere and the remainder is eventually decomposed by the soil microorganisms. A later section in this chapter discusses volatile organics in greater detail. Decomposition of animal and vegetable oils should occur at a higher rate than petroleum since the materials are more readily degradable by the soil organisms.

Kincannon (35) applied oil rich sludges (petroleum base) to soils and monitored the rate of oil loss. The control plots received no nutrient fertilizers (N,P,K) and an average oil loss of 0.52 lb/month per cubic foot of soil was observed. Some combination of volatilization and microbial activity was responsible for the loss. The addition of commercial fertilizers doubled the rate of oil loss; with most of the loss occuring in the warm months, indicating that microbial action was the major pathway. On an annual basis the rate of oil loss was 3.3 tons per acre on the control plots and 67 tons per acre on the fertilized and cultivated plots (35). Overcash (34) indicates that microbial decomposition could remove up to 96 tons of soybean oil per acre per year.

The addition of oil to the soil plant matrix changes the carbon to nitrogen (C:N) ratio significantly. The addition of extra nitrogen and other micro nutrients is necessary for the microbial reactions to proceed at acceptable rates. Kincannon (35) added about 1000 lb of N and 200 lb of P_2O_5 per acre to achieve the maximum oil loss rates discussed above. Overcash (34) cites work recommending 0.005% N and 0.002% P on a soil weight basis to achieve maximum degradation of some oils.

Oil can also have a negative effect on the germination of seed when applied to an agricultural land treatment site. If the oil has a significant volatile faction it should be applied well before or well after the planting and germination period. The impact on germination and yield of the vegetation is more significant than the impact on the soil system. An oil level of about 1% of soil weight seems to be the threshold for reduced yields and at levels of 1.5-2% the reduction in yield often exceeds 50% (34). These effects occur with a fresh application of oil prior to the loss of the hydrocarbons.

A soil depth of about 6 in. should be used to determine acceptable oil or grease loading rates. The maximum single dose that can be applied should be determined then the insitu decomposition rates will determine the interval between applications. If there is no surface vegetation a loading equal to 2-4% of the soil weight in the top six inches might be acceptable. If there is a crop a single dose of 1-2% would significantly reduce yield. A warm weather (soil temperatures 50° - 60°F) decomposition rate of 0.2-0.4% oil, of soil weight per month is suggested (34). Short term on-site tests are recommended for final system design with a particular oil. Table 3-8 summarizes the oil tolerance for a range of crops.

Table 3.8: Oil Tolerance for Selected Crops (34)

Crop Type	Single Oil Application
yams, carrots, rape, lawn-grasses, sugar beets	<0.5% of soil weight (<5 ton oil/acre)
ryegrass, oat, barley, corn, wheat, beans, soybeans, tomato	<1.5% of soil weight (<15 tons of oil/acre)
red clover, peas, cotton, potato, sorghum	<3% soil weight (<30 tons of oil/acre)
perennial grasses, coastal Bermuda grass, trees, plantain	>3% soil weight (>30 tons of oil/acre)

Example 3-2: Oil Degradation Rate

Condition: Industrial waste with 1% vegetable oil, soil degradation rate 0.15% of soil weight per month. Corn is intended crop with acceptable oil tolerance at 0.5% of soil weight. Soil density at surface 90 lb/ft^3. Find waste loading per acre.

Solution:

Soil in top 6 in. = (90)(0.5)(43,560) = 1,960,200 lb/acre

Allowable oil load = (0.005)(1,960,200) = 9801 lb/acre.

$$\text{Waste application} = \frac{9801 \text{ lb/acre}}{(0.01)(2000) \text{ lb/ton}} = 490 \text{ tons/acre}$$

At 0.15% per month the oil would be degraded in:

$$\frac{0.5\%\ \text{applied}}{0.15\%/\text{month degraded}} = 3.3\ \text{months}$$

METALS

Trace metals removal in the soil is a complex process involving the mechanisms of adsorption, precipitation, ion exchange, and complexation. Because adsorption of most trace elements occurs on the surfaces of clay minerals, metal oxides, and organic matter, fine textured and organic soils have a greater adsorption capacity for trace elements than sandy soils. The SR systems are the most effective for metals removal because of the finer textured soils and the greater opportunity for contact and adsorption. Rapid infiltration is also quite effective but the travel distance in the soil will have to be longer due to the higher hydraulic loadings and the coarser textured soils. Overland flow systems allow minimal contact with the soil and typically remove between 60 and 90% depending on the hydraulic loading and specific metal (6).

In general, metals are present in typical municipal wastewaters in very low concentrations. As shown on Table 3-9 the typical concentrations in raw sewage are below the requirements for drinking and irrigation waters.

Table 3.9: Metals in Municipal Wastewater and Allowable Concentrations in Drinking and Irrigation Waters

Element	Raw Sewage (mg/ℓ)	Drinking Water (mg/ℓ)	. . . .Irrigation (mg/ℓ) . . . . 20 yr*	Continuous**
Cadmium	0.004-0.14	0.01	0.05	0.005
Chromium	0.02-0.70	0.05	20	5.0
Lead	0.05-1.27	0.05	20	5.0
Zinc	0.05-1.27	0.05	20	5.0

*For fine textured soils only, normal irrigation practice for 20 years.
**For any soil, normal irrigation practice, no time limit.

Wastewater treatment by activated sludge and similar processes tends to concentrate these metals in the sludge fraction. The land application of these sludges often then raises concerns with respect to metal contamination; the land application of the liquid effluents from these processes should not therefore be a problem.

Metal Limits

The major concern with respect to metals is the potential for accumulation in the soil profile and then subsequent translocation, via crops or animals, through the food chain to man. The metals of greatest concern are lead (Pb), zinc (Zn), copper (Cu), nickel (Ni) and cadmium (Cd) (6). Most crops do not accumulate Pb but there is some concern with respect to ingestion by animals grazing on forages or soil to which sludge has been applied. In general, Zn, Cu, and Ni will be toxic to crops before their concentration in plant tissues reaches a level that poses a problem to human or animal health. Cadmium is the greatest concern since the concentration of concern to human health are far below those that would produce toxic effects in the crops. The U.S. Department of Agriculture and the U.S. Environmental Protection Agency developed guidelines for cumulative metals additions to agricultural crop land (36). No adverse effects of any kind would be expected at these rates.

Table 3.10: USEPA Recommended Cumulative Limits for Metals Applied to Agricultural Crop Land (36)

	........Soil Cation Capacity (meq/100 g).........		
	5	5-15	>15
	Sands	Sandy Loams	Silt Loams
Metals	..........Cumulative Loading (lb/acre)..........		
Pb	500	1,000	2,000
Zn	250	500	1,000
Cu	125	250	500
Ni	125	250	500
Cd	4.4	8.9	17.8

The limits were developed for sludge applications but should be applicable for wastewaters, particularly industrial types. These limits assume that the soil pH is maintained at 6.5 or above during and after application of the waste materials. These values may limit the useful life of a site if a particular metal is present in high concentrations. Such constraints are not likely to apply for typical municipal effluents.

Example 3-3: Cadmium Loadings

Conditions: Slow rate land treatment site, wastewater application 8 ft per year. Cd concentration in wastewater 0.01 mg/L, silt loam soil. Find the time limit for cadmium applications.

Solution:

$$\left(\frac{8\text{ft}}{\text{yr}}\right)(43{,}560\ \text{ft}^2/\text{acre})(7.48\ \text{gal/ft}^3) = 2{,}606{,}630\ \frac{\text{gal}}{\text{yr.}}$$

$$(0.01\ \text{mg/L})(8.34)(2.6066) = 0.22\ \text{lb Cd/yr}$$

Cumulative limit = 17.8 lb/acre (Table 3-10)

$$\frac{17.8\ \text{lb/acre}}{0.22} = 81\ \text{years}$$

Since the limits incorporate very significant safety factors it is likely that even higher loadings will be possible before any adverse effects occur. This is particularly true for effluent applications that will take greater advantage of the adsorptive capacity of the entire soil profile and not be limited to the near surface zone that interacts with sludges. The values in Table 3-10 are not applicable to RI or OF systems.

Soil and Crop Removals

It is not possible to predict the total renovative capacity of a site with simple ion exchange or soil adsorption theories. Although the metals are accumulated in the soil profile the accumulation does not seem to be continuously available for crop uptake. Work by Hinesly (38) (39) and others with sludges demonstrate that the metal uptake by crops in a given year is more dependent on the concentration of metals in the sludge most recently applied and not on the total accumulation of metals in the soil.

The capability for metal uptake varies with the type of crop grown. Swiss Chard and other leafy vegetables take up more metals than other types of vegetation. Metals tend to accumulate in the liver and kidney tissue of animals grazing on the land treatment site or if fed harvested products. A number of studies with domestic and indigenous animals do not show adverse effects. Tests done on a mixed group of 60 Hereford and Angus steers that graze directly on the grasses at the Melbourne, Australia site showed that: "the concentration of zinc, cadmium and nickel found in the liver and kidney tissues of this group of cattle are within the expected normal range of mammalian tissue" (40). Anthony (41) has reported on metals in bone, kidney and liver tissue in mice and rabbits indigenous to the Pennsylvania State University land treatment site and no adverse responses were noted.

The average metal concentrations in the shallow groundwater beneath the Hollister, CA rapid infiltration site are shown in Table 3-11. After 33 years of operation the concentration of cadmium, chromium and cobalt were not significantly different from normal off-site groundwater quality. The concentrations of the other metals were higher than background.

Table 3.11: Trace Metals in Groundwater Under Hollister, CA, Rapid Infiltration Site (42)

Metal	Average Concentration (mg/ℓ)
Cadmium	0.028
Cobalt	0.010
Chromium	0.014
Copper	0.038
Iron	0.36
Manganese	0.96
Nickel	0.13
Lead	0.09
Zinc	0.081

The metal concentrations in the upper foot of soil at the Hollister, CA site are still below or near the low end of the range cited as typical for agricultural soils after 33 years of operation.

In OF the major mechanisms responsible for trace element

removal include sorption of clay colloids and organic matter at the soil surface layer, precipitation as insoluble hydroxy complexes, and formation of organometallic complexes with the organic matter at the slope surface. The largest proportion of the heavy metals accumulate in the biomass on the soil surface and close to the point of effluent application.

In summary, it is unlikely that metals will be the LDP for land treatment of municipal wastewaters. It is possible, that a metal could be the LDP for an industrial land treatment operations, and it is probable that metals will be LDP for sludge (See Chapter 16).

NITROGEN

The removal of nitrogen in land treatment systems is complex and dynamic due to the many potential forms of nitrogen (N_2, organic N, NH_3, NH_4, NO_2, NO_3) and the relative ease of changing from one oxidation state to the next. The nitrogen in typical untreated municipal wastewater is usually present as organic nitrogen (40%) and ammonia/ammonium ions (60%). Activated sludge and other high rate biological processes can be specifically designed to nitrify all the nitrogen remaining in the effluent but typically only a portion is nitrified and the major fraction in most system effluents is still ammonia.

It is important in the design of all three land treatment concepts to identify the total concentration of nitrogen in the wastewater to be treated and the specific forms that are expected. Experience with all three processes demonstrates that the less oxidized the nitrogen is entering the system the more effective will be the retention and overall N removal (6).

Soil Responses

The soil plant system provides a number of interrelated responses to wastewater nitrogen. The organic N fraction, usually associated with particulate matter is entrapped or filtered out of the liquid stream. The ammonia fraction can be lost by volatilization, taken up by the crop or adsorped by the clay minerals in the soil. The latter is a renewable process since the soil microbes oxidize the retained ammonium to nitrate and restore the adsorptive capacity of the soil. Nitrate can be taken up by the vegetation, or converted to nitrogen gas via denitrification in an anaerobic zones and lost to the atmosphere. The decomposition of the organic nitrogen tied up in particulate matter proceeds more slowly. This is more critical for sludge application systems where the solids fraction is a very significant part of the total application. In those cases a sequence of mineralization rates are used to account for the annual nitrogen contribution from previous applications of sludge (44). This could also be a factor for high solids content industrial wastewaters but the design for typical municipal effluents does not include this refinement.

Nitrification (conversion of ammonium ions to nitrate) is very effective in all three of the basic land treatment concepts. Under favorable conditions nitrification ranging from 5 to 50 mg/L N per day is possible. Assuming that these reactions are occurring with the adsorbed ammonium ions in the top 4 in. of soils, means that up to 60 lb of ammonia nitrogen per acre can be converted to nitrate each day. At a typical wastewater concentration of 20 mg/L NH_4-N, up to 12 in. of wastewater per day could theoretically be applied if the soil could be maintained in an aerobic condition.

The maintenance and/or restoration of the necessary aerobic conditions is the reason for the short application periods and cyclic operations typically used in land treatment systems. In RI for example it is believed that the adsorption sites are saturated with ammonium ions during the early part of the application period. Then aerobic conditions are restored during the rest period and the soil microbes convert that ammonium to nitrate. At the next application cycle the "new" ammonia is again adsorbed, some of the nitrate moves downward with the percolate but a significant part of the nitrate is denitrified when anaerobic conditions occur and is then lost to the atmosphere. Denitrifying bacteria are common soil organisms and the occurrence of anaerobic conditions at least at microsites can be expected with SR and OF systems as well as RI.

Nitrification is a conversion process not a removal process for nitrogen. Denitrification, volatilization and crop uptake are the only true removal pathways available. Crop uptake is the major factor for most slow rate systems with additional contributions from denitrification and ammonia volatilization depending on site conditions and wastewater type. In RI, denitrification is the only important mechanism. For OF crop uptake and denitrification both contribute. Crop uptake of nitrogen is discussed in detail in Chapter 5 and in the process design chapters. Nitrogen removal data for typical SR, RI, and OF systems are shown on Table 3-12.

Table 3.12: Nitrogen Removal of Typical Land Treatment Systems (6)

Process/Location	Concentration in Wastewater (mg/ℓ)	Concentration in Renovated Water (mg/ℓ)
SR		
Dickinson, ND	12	3.9
Hanover, NH	28	7.3
Roswell, NM	66	10.7
San Angelo, TX	35	6.1
RI		
Calumet, MI	24	7
Ft. Devens, MA	50	20
Hollister, CA	40	3
Phoenix, AZ	27	10
OF		
Ada, OK, raw wastewater	34	7
primary effluent	19	5
secondary effluent	16	8
Easley, SC pond effluent	7	2
Utica, MS pond effluent	20	7

Nitrates

The health issue of concern for nitrogen is excess concentration of nitrate in drinking waters for infants under 6 months of age. The U.S. primary drinking water standard for nitrate (as N) is set at 10 mg/L for this reason. The pathway of concern in SR and RI systems is conversion of wastewater nitrogen to nitrate and then percolation to drinking water aquifers. When potable aquifers are involved, the current guidance requires that all drinking water standards be met at the project boundary. As a result, nitrogen often becomes the LDP for SR systems because of its relatively high concentration as compared to the other drinking water parameters. Chapter 10 presents complete design details for nitrogen removal.

There are a number of safety factors inherent in the approach that insure a conservative design. The procedure assumes that all of the applied nitrogen will appear as nitrate (ie. complete nitrification) and within the same time period assumed for application (no lag or mineralization) and there is no credit for mixing or dispersion with the insitu groundwater.

Design Factors

The mass balance for RI systems would not typically include a component for crop uptake. The percolate N concentration is not a concern for OF systems since the percolate volume is generally assumed to be negligible. As indicated previously, sludge application systems also include a mineralization factor to account for previous organic nitrogen deposits. There are four potential situations where such a factor might be included in the balance for SR and OF systems also:

- Industrial wastewaters with high solids (significant N content in solids) SR or OF.
- Grass covered systems where the grass is cut but not removed. SR or OF.
- Pasture systems with animal grazing and animal manure left on site. SR.

- Sludge or manure added to an SR site as supplemental fertilizer.

Organic Nitrogen

Mineralization rates, developed for wastewater sludge are given in Table 3-13. The values are the percent of the organic nitrogen present that is mineralized (ie. converted to inorganic forms, ammonia, nitrate, etc.) in a given year. For example, 40% of the organic nitrogen present in raw sludge would be mineralized during the first year, 20% of the organic nitrogen remaining after the first year would be mineralized during the second year, etc.

Table 3.13: Mineralization Rates for Organic Nitrogen in Wastewater Sludges (44)

Time After Sludge Application (yr)	...............Mineralization Rate (%)...............		
	Raw Sludge	Anaerobically Digested	Composted Sludge
First	40	20	10
Second	20	10	5
Third	10	5	3
Fourth	5	3	3
Fifth	3	3	3
Sixth	3	3	3
Seventh	3	3	3
Eighth	3	3	3
Ninth	3	3	3
Tenth	3	3	3

The mineralization rate appears to be related to initial organic nitrogen concentration (43), so the higher the organic N content is the higher the mineralization rate will be. Easily degraded industrial solids would be comparable to raw sludge. Industrial solids with a high percentage of relatively stable humic substances should be similar to composted sludges. Animal manures would be similar to digested sludges and it would be conservative to also assume that grass cuttings and vegetative litter would decay at the same rates as digested sludges. Another consideration is necessary for animal manures to account for volatization of the ammonia faction. Since the manure is deposited on the surface, essentially all of the ammonia content will be lost to the atmosphere, leaving the organic fraction to be mineralized. If the manure is similar in composition to human wastes about 50% of the nitrogen will be present in the ammonia form with the remainder as organic nitrogen. The examples below illustrate the use of the factors in Table 3-13 for two possible situations.

Example 3-4: Nitrogen Cycling in Greenbelts

Conditions: Slow rate land treatment site, used as a greenbelt parkway. The grasses are cut but not removed from the site. At the loading rates used the grass species will take up about 300 lb/acre of nitrogen per year. Find the nitrogen contribution from on-site decay of the cut grass.

Solution: Most conservative assumption is to use digested sludge rates from Table 3-13; and assume that all nitrogen is in the organic form.

1. First year: (300 lb/acre)(0.20) = 60 lb/acre
2. Second year: Second year cutting (300)(0.20) = 60 lb/acre
   Residue from prior years =
   (300-60)(0.10) = 24
   Total = 84 lb/acre
3. Third year: Third year cuttings = (300)(0.20) = 60
   Residue from second year = (300-60)(0.10) = 24
   Residue from first year =(300-84)(0.05) = 11
   95 lb/acre
4. Fourth year: Fourth year cuttings =
   (300)(0.20) = 60
   Residue from third year =
   (300-60)(0.10) = 24
   Residue from second year =
   (300-84)(0.05) = 11
   Residue from first year =
   (300-95)(0.03) = 6
   Total = 101 lb/acre

5. Fifth year: Fifth year cuttings = 60

Residue from fourth year = 24

Residue from third year = 11

Residue from second year = 6

Residue from first year =

(300-101)(0.03) = 6

Total = 107 lb/acre

6. The amount of nitrogen contributed becomes relatively stable after the third or fourth year and increase only slightly thereafter. In this case it should be assumed that about 100 lb/acre of nitrogen is returned to the soil system each year from the cut grass. For this example that would be about 33% of the nitrogen that was originally removed by the grass. That is a very significant amount and should be included in the mass balance for design.

Example 3-5: Nitrogen Cycling in Grazed Pastures

Conditions. This example assumes similar grasses with similar nitrogen uptake but on a pasture with grazing cattle. Find the nitrogen contribution from decay of animal manures on site.

Solution: To be conservative assume that all of the grass is consumed and that all of the nitrogen present in the grass is present in the manure, all of which is deposited on site.

1. Annual organic N = Total N - ammonia volatilization

= 300 lb/acre - (0.50)(300) = 150 lb/acre

2. Use digested sludge mineralization rates from Table 3-13.

First year: (150)(0.20) = 30 lb/acre

Second year: 30 + (150-30)(0.10) = 42 lb/acre

Third year: 30 + 12 + (150-42) (0.05) = 47 lb/acre

etc.

3. Therefore, the animal manure will return about 50 lb/acre of nitrogen to the soil system each year after equilibrium is

reached. That is only 17% of the wastewater nitrogen contained in the grass but is still significant and should be included in the mass balance (if the manure (or sludges) were immediately plowed in and incorporated with the soil the 50% credit for volatilization would not occur).

These two examples illustrate the critical importance of knowing which form the nitrogen is in (NH_4, NH_3, Organic N, etc.) when it is applied to the land treatment site. This is particularly important if elaborate pretreatment is provided since the nitrogen may not then be in the simple combination of organic N and ammonia that was originally present in the untreated municipal sewage. In addition, any nitrogen losses that occur during preapplication treatment or storage should be considered. Facultative lagoons or storage ponds can remove up to 95% of the contained nitrogen under ideal conditions, as described in Chapter 8. Such losses are especially significant when nitrogen is the LDP for design since any reduction in nitrogen during preapplication treatment or storage will then proportionally reduce the size and therefore the cost of the final land treatment site.

PHOSPHORUS

The presence of phosphorus (P) in water supplies is not known to have any health significance but phosphorus is considered to be the limiting factor for eutrophication of surface waters so its removal from wastewater is a concern to many. Phosphorus is present in municipal wastewater as orthophosphate, polyphosphate and organic phosphates. The orthophosphates are immediately available for biological reactions in soil systems. The necessary hydrolysis of the polyphosphates proceeds very slowly in typical soils so these forms are not as readily available. Industrial wastewaters may contain a significant fraction of organic phosphorus but typical municipal wastewaters do not.

Removal Mechanisms

Phosphorus removal in land treatment systems occurs through plant uptake, biological, chemical and/or physical processes. As described in the previous section nitrogen removal is almost entirely dependent on biological processes so the removal capacity can be maintained continuously and/or restored by proper design and management. In contrast, phosphorus removal depends to a very significant degree on chemical reactions which are not necessarily renewable. As a result the retention capacity for phosphorus in a deposit of soil will be decreased, in time, but not exhausted. At typical SR systems for example it is estimated that a one foot depth of soil will become saturated with phosphorus every 10 years. Removal of phosphorus will be almost complete during this period and phosphorus should not be a problem until all of the soil is utilized and the percolate emerges and/or is extracted for discharge to surface waters.

It is unlikely that phosphorus in municipal wastewaters would be the LDP for process design of the land treatment system. However, in those situations where P removal is required by the regulatory agencies the design life of the site may be controlled by phosphorus considerations. The most critical case might be a site with relatively clean sands and underdrains at a shallow depth discharging to a sensitive surface water. The P removal capacity in such a situation might range from 20 to 60 years depending on the soil type, wastewater characteristics, and loading rates.

The phosphorus removals observed at typical SR and RI systems are summarized on Table 3-14. Crop uptake contributes to P removal at SR systems but in both SR and RI the major removal occurs in the soil. The phosphorus is removed by adsorption/precipitation reactions when clay, oxides of iron and aluminum, and calcareous substances are present. The P removal increases with increasing clay content and with increasing detention time in the soil. The percolate phosphorus values listed in Table 3-14 for SR systems are close to the background values for natural groundwater in those locations.

Table 3.14: Typical Phosphorus Responses (6, 45)

Location	Soil Type	Travel Distance (ft)	Percolate Phosphorus (mg/ℓ)
SR			
Hanover, NH	sandy loam	5	0.05-0.06
Muskegon, MI	loamy sand	5	0.03-0.05
Tallahassee, FL	fine sand	4	0.1
Penn State (forest)	silt loam	4	0.08
Helen, GA (forest)	sandy loam	4	0.17
RI			
Hollister, CA	gravelly sand	22	7.4
Phoenix, AZ	gravelly sand	30	4.5
Ft. Devens, MA	gravelly sand	5	9.0
Calumet, MI	gravelly sand	30	0.1
Boulder, CO	gravelly sand	10	0.02-4.5
Lake George, NY	sand	3	1
		600	0.014
Vineland, NJ	sand	30	1.5
		400	0.27

Rapid Infiltration

There is no crop uptake in RI systems and the nature of the soils and high hydraulic rates typically used require greater travel distances in the soil for effective P removal. Data from several of the RI systems in Table 3-14 indicate a phosphorus concentration approaching background levels after several hundred feet of travel away from the site. Most of these systems (Vineland, Lake George, Calumet, Ft. Devens) have been in operation for several decades.

An equation to predict phosphorus removal at RI sites has been developed from data collected at a number of operating systems (6):

$$P_x = P_o e^{-kt} \qquad (3\text{-}3)$$

where P_x = total phosphorus concentration at a distance x along the percolate flow path, mg/L

P_o = total phosphorus concentration in the applied wastewater, mg/L

k = instantaneous rate constant and equals 0.002 h^{-1} at neutral pH

t = detention time = $X\theta/I$, (h)

where X = distance along the flow path, in.

θ = volumetric water content, in.^{3}/in.^{3}, use 0.4

I = infiltration rate during system operation, in./h (use basin test results, 20% of cylinder infiltration results, or horizontal conductivity for horizontal flow)

Because the minimum phosphorus precipitation rate occurs at neutral pH, this equation can be used to conservatively estimate phosphorus removal. If the calculated phosphorus concentration is an acceptable value, phosphorus concentrations from an operating RI system should be well within any required limits. However, if the calculated phosphorus concentration at a distance x exceeds required values, phosphorus adsorption tests should be performed. This test measures the ability of a specific soil to remove phosphorus and is described in Chapter 7.

Example 3-6: Phosphorus Removal

Conditions: Consider a site where wastewater percolates through the soil to the groundwater table, which is 49 ft below the soil surface. The initial phosphorus concentration is 10 mg/L and the basin infiltration test rate is 16 in./d.

Solution:

1. The phosphorus concentration at the 49 foot depth should be:

$$(10 \text{ mg/L})\ e^{-0.002h^{-1}\left(\frac{49 \text{ ft} \times 0.40}{1.33 \text{ ft/d}}\right)} = 4.9 \text{ mg/L}$$

2. If the movement is then predominantly horizontal, with the

renovated water seeping into a creek 650 ft from the infiltration site, and the horizontal hydraulic conductivity is 4 ft/d, the phosphorus concentration in the seepage should be less than:

$$(4.9 \text{ mg/L})e^{-0.002h^{-1}} \left(\frac{650 \text{ ft x } 0.40}{4 \text{ ft/d}}\right)\left(\frac{24h}{d}\right) = 0.2 \text{ mg/L}$$

Since equation 3-3 was developed for the relatively coarse textured soils typically used in RI systems it is not appropriate for predicting phosphorus responses at SR systems since it will overestimate the travel distance required in these more reactive soils. If P removal is required in an SR system, phosphorus adsorption tests, with the specific soils to be used, should be performed.

Overland Flow

The opportunities for contact between the wastewater and the soil are limited in OF systems and as a result P removals typically range from 40 to 60%. P removal in overland flow can be improved by chemical addition and precipitation on the treatment slope. At Ada, OK alum additions (Al to P mole ratio 2:1) reduced P in the runoff to 1.0 mg/L (6). At Utica, MS mass removals ranged between 65 and 90% with alum as compared to less than 50% without alum (45).

Typical municipal wastewaters will have between 5-20 mg/L of total phosphorus. Industrial wastewaters can have much higher concentrations particularly from the fertilizer and detergent industries (34). Food processing operations can also have high phosphate effluents. Some typical values are:

Dairy products = 9-210 mg/L PO_4

Grain milling = 5-100 mg/L PO_4

Cattle feed lots = 60-1500 mg/L PO_4

Phosphorus could be the LDP for such wastewaters in determining the design or operational life of the system.

Example 3-7: Phosphorus Adsorption

Conditions: Assume a silty loam soil, adsorption tests indicate useful capacity for phosphorus equal to 3000 lb/acre of P per foot of depth. Site to be grass covered, grass to be harvested and removed. P uptake of grass equal to 30 lb/acre·yr of P. Operational life of factory and treatment site equal to 30 years. Phosphorus in wastewater at 60 mg/L. Find the average annual wastewater loading.

Solution: Site investigation indicates top 6 ft of soil profile will be usable for P removal over the life of the site.

1. Soil contribution = (3000 lb/acre·ft)(6 ft)
   = 18,000 lb/acre

2. Crop contribution = (30 lb/acre·yr)(30 yr) = 900 lb/acre

   Total P capacity = 18,900 lb/acre

   Average annual loading = $\frac{18,900}{30}$ = 630 lb/yr

3. Wastewater = (60 mg/L) (8.34)(Q) = 630 lb/yr

   Q = 1,260,000 gallons/acre.yr

   = 3.86 ft of wastewater per yr, for 30 years.

INORGANIC TRACE ELEMENTS AND SALTS

This category refers to nonmetallic elements such as boron, selenium, arsenic, sodium, sulfur, potassium, and the compounds, oxides and salts formed from these materials. The major impact of these materials in land treatment systems is on the vegetative component and on permeability of certain clays due to high sodium in the wastewater. Some, such as potassium, are essential plant nutrients, others serve as micronutrients at low concentration but can be toxic to plants at high levels. At the concentrations found in typical municipal wastewaters none of these materials are likely to be the LDP for process design. One exception, discussed below, might be high salinity in wastewaters applied to the land in relatively arid climates.

Boron

Boron is an essential micronutrient for plants but becomes toxic at relatively low concentrations (<1 mg/L) for sensitive plants. The soil has some adsorptive capacity for boron if aluminum and iron oxides are present. The reactions are similar to those described previously for phosphorus but the capacity for boron is low. A conservative design approach would assume that any boron not taken up by the plants would be available for percolation to the groundwater. Plant uptake of boron in corn silage of about 0.005 lb/acre·yr and in alfalfa of 0.81-1.6 lb/acre·yr have been reported (34). At the SR land treatment site at Mesa, AZ the applied municipal effluent had 0.44 mg/L boron and the groundwater beneath the site contained 0.6 mg/L. At another SR operation at Camarillo, CA the wastewater boron was 0.85 mg/L and the groundwater was up to 1.14 immediately beneath the site. The slight increase in both cases is probably due to water losses from evapotranspiration. Table 3-15 lists suggested vegetation types with respect to their boron tolerance.

Table 3.15: Boron Tolerance of Crops (46)

I Tolerant	II Semitolerant	III Sensitive
Alfalfa	Barley	Fruit crops
Cotton	Corn	Nut tree
Sugar beet	Milo	
Sweet clover	Oats	
Turnip	Tobacco	
	Wheat	

Overcash (34) has suggested that industrial effluents with 2-4 mg/L boron could be successfully applied to the crops in category I in Table 3-15, 1-2 mg/L boron for category II and less than 1 mg/L for category III. Boron is therefore not the LDP for process design but may be the determinant on which crop to select.

Both RI and OF systems will be less effective for boron removal than SR because of the same factors discussed for phosphorus. Early injection experiments at Orange County, CA for example injected effluent with 0.95 mg/L boron. After 545 ft travel in the soil the concentration was still 0.84 mg/L (46).

Selenium

Selenium is a micronutrient for animals but is non essential for plants. However, in high concentrations it is toxic to animals and many plants can accumulate selenium to those levels without any apparent effect on the crop. Plants containing 4 to 5 ppm selenium are considered toxic to animals (46). Selenium can be adsorbed weakly by the hydrous iron oxides in soils and this is of more concern in the southeastern U.S. where soils have very high iron oxide contents. Selenium is not likely to be the LDP for land treatment design for either municipal or industrial systems. However, selenium is included on EPA's list of priority pollutants and if concentrations greater than 0.01 mg/L are expected in industrial effluents it may be necessary to avoid SR or OF land treatment options because of potential long-term adverse impacts if the harvested crops enter the human food chain.

Arsenic

Arsenic is non essential for all life forms. In significant concentrations it can be moderately toxic to plants and very toxic to animals. The food chain is protected at land treatment sites since the crops should show adverse effects from arsenic before hazardous levels were reached in the edible portions. Arsenic is removed in the soil system by adsorption by the soil colloids with clay and the iron and aluminum oxides performing essentially the same function as described previously for phosphorus removal. In general, arsenic will not be the LDP land treatment of municipal wastewaters. Poultry manure with 15-30 ppm arsenic have been applied for up to 20 years (0.2-0.4 lb As/acre·yr) without any adverse effects on either alfalfa or clover (47). Field tests are recommended for industrial effluents with high arsenic concentrations to develop criteria for loading rates and vegetation to be used in a specific location.

Sodium

Sodium is typically present in all wastewaters. There are no primary drinking water requirements for sodium but it has been suggested that human consumption of high levels of sodium is related to heart disease. Sodium and calcium can be directly toxic to plants but most often their influence on soil salinity or alkalinity is the most important problem. Growth of sensitive plants becomes impaired where the salt content of the soil exceeds 0.1%. Salinity has a direct bearing on the osmotic pressure of the soil solution which controls the ability of the plant to absorb water. Adverse crop effects can also occur from sprinkler operations using water containing significant sodium or chlorine. Leaves can absorb both elements rapidly and their accumulation on the leaves in arid regions can result in toxicity problems (46). Sodium is not permanently removed in the soil but is rather involved in the cation exchange processes. These reactions are similar to those occurring in water softening processes and involve sodium, magnesium, and calcium in the system.

In some cases, where there is an excess of sodium with respect to calcium and magnesium in the water, applied to high clay content soils, there can be an adverse effect on soil structure. The deflocculation and swelling of clay particles can significantly reduce the hydraulic capacity of the soil. The relationship between sodium, calcium, and magnesium is expressed as the Sodium Adsorption Ratio (SAR):

$$SAR = \frac{Na}{\left(\frac{Ca + Mg}{2}\right)^{\frac{1}{2}}} \qquad (3\text{-}4)$$

Concentrations of Na, Ca and Mg expressed as milliequivalents per liter.

Example 3-8: Calculate SAR

Conditions:

Atomic Wt: Na 22.99, Ca 40.08, Mg 24.32. Find SAR for a wastewater with: Na = 37.9 mg/L, Ca = 10.8 mg/L, Mg = 3.8 mg/L.

Solution:

$$\text{Meq (Na)} = \frac{(37.9)(1)}{(22.99)} = 1.65$$

$$\text{Meq (Ca)} = \frac{(2)(10.8)}{40.08} = 0.54$$

$$\text{Meq (Mg)} = \frac{(2)\ (3.8)}{(24.32)} = 0.32$$

$$\text{SAR} = \frac{1.65}{\left(\frac{0.54 + 0.31}{2}\right)^{\frac{1}{2}}} = 2.53$$

An SAR of 10 or less should be acceptable on soils, with significant clay content (15% clay). Soils with little clay, or non swelling clays can tolerate an SAR up to 20. It is unlikely that any problems of this type will occur with application of municipal effluents in any climate since the SAR seldom exceeds a value of 5 to 8. Industrial wastewaters may be more of a concern. The water from ion exchange softening could have an SAR of 50, and some food processing effluents range from about 30 to over 90.

The common remedial measure for SAR induced soil swelling or permeability loss is the surface application of gypsum or another inexpensive source of calcium. The addition of water allows the calcium to leach into the soil to exchange with the sodium. An additional volume of water is then required to leach out the salt solution.

Salinity problems are of most concern in the arid regions since the wastewater to be applied may already have a high salt content; the concentration will be further increased due to evapotranspiration on the site, and because system design is usually based on satisfying the minimal water needs of the crop to be grown. The combination of these factors will result in the rapid build-up of salts in the soils unless some mitigation efforts are applied. The standard approach is to determine crop water needs and then add to that a leaching requirement (LR) to insure that an adequate volume of water passes through the root zone to control salts at a specified level. The LR can be determined if the salinity or electrical (EC) of the irrigation water and the required EC in the drainage water to protect a specific crop are known (47). The salt content of irrigation waters is often expressed as mg/L, and can be converted to conductivity terms (mmho/cm) by dividing mg/L by 640. The equation to estimate the LR is:

$$LR = \frac{(EC)_I}{(EC)_D} \times 100 \qquad (3\text{-}5)$$

where LR = leaching requirement as a percent.

$(EC)_I$ = average conductivity of irrigation water (including natural precipitation), mmho/cm

$(EC)_D$ = required conductivity in drainage water to protect the crops, mmho/cm.

Values of $(EC)_D$ for typical crops with no yield reduction (see reference (47) for additional detail) are presented in Table 3-16.

Table 3.16: Values of EC_D for Crops with no Yield Reduction (49)

Crop	Electrical Conductivity (EC_D), mmho/cm
Bermuda grass	13
Barley	12
Sugar beets	10
Cotton	10
Wheat	7
Tall fescue	7
Soybeans	5
Corn	5
Alfalfa	4
Orchard grass	3

The total water application required is then:

$$LW = \frac{CU}{1 - \frac{LR}{100}} \qquad (3\text{-}6)$$

where LW = required water application, in.

CU = consumptive use of water by the crop between water applications, in.

LR = leaching requirement (as a percent)

Example 3-9: Leaching Requirement

Conditions: Given a wastewater at 800 mg/L salinity and corn, with $(EC)_D = 5 \times 10^3$ mmho/cm, consumptive use between irrigations = 3 in. Find the total water requirement.

Solution: Conductivity of wastewater =

$$\frac{800}{640} = 1.25 \times 10^3 \ \frac{\text{mmho}}{\text{cm}}$$

$$LR = \frac{1.25}{5} \times 100 = 25\%$$

$$LW = \frac{3}{1-0.25} = 4 \text{ in.}$$

Salinity problems and leaching requirements are not to be expected for land treatment systems in the more humid portions of the U.S. since natural precipitation is higher and higher hydraulic loadings are typically used to minimize the land area required.

Sulfur

Sulfur typically is present in most wastewaters either in the sulfate or sulfite form. Sulfate is not strongly retained by the soils but is usually present in the soil solution. Sulfates are not typically present in high enough concentrations in municipal wastewaters to be a concern for design of land treatment systems. Drinking water standards limit sulfate to 250 mg/L, irrigation standards recommend 200-600 mg/L depending on the type of vegetation. Industrial wastewaters from sugar refining, petroleum refining, and

Kraft process paper mills might all have sulfate or sulfite concentrations requiring special concern. Crop uptake can account for some sulfur removal, Table 3-17 summarizes typical values for several crops.

Table 3.17: Sulfur Uptake by Selected Crops (34)

Crop	Harvested	Sulfur (lb/acre)
Corn	200 bu/ac	44
Wheat	83 bu/ac	22
Barley	100 bu/ac	25
Alfalfa	6 ton/ac	30
Clover	4 ton/ac	18
Coastal Bermuda grass	10 ton/ac	45
Orchard grass	7 ton/ac	50
Soybeans (beans only)	50 bu/ac	10
Cotton	2.5 bale/ac	23

The design approach is similar to that described previously for nitrogen. It is prudent to assume that most of the sulfur compounds applied to the land will be mineralized to sulfate. The 250 mg/L standard for sulfate would then apply at the project boundary when drinking water aquifers are involved. The major permanent removal pathway is to the crop and values in Table 3-17 can be used for estimating purposes. If industrial wastes have particularly high organic contents there may be additional immobilization of sulfur. It is recommended that specific tests be run for industrial wastewaters of concern to determine the soil adsorptive capacity and the potential for organic immobilization of sulfur.

Potassium

Potassium (K) is an essential nutrient for plant growth and is typically found in low to moderate concentrations (15 mg/L in typical municipal effluents). Potassium is in the same cationic group as Na, Mg and Ca. It interacts with the soil cationic exchange processes but is not strongly held and not accumulated in the soil system.

The concern in land treatment is not with respect to removal of potassium but rather with respect to the potential for a deficiency and this deficiency in turn affecting the removal of N or P by the crop. Each crop has a specific ratio of nutrient requirements (eg: N:P:K) to optimize growth or from a treatment point of view to optimize removal of N or P. If the proper balance is not maintained, removal of N or P will be less than expected. In general there is an excess of nitrogen in municipal wastewaters with respect to potassium for most agricultural and forage crops:

Typical Municipal Wastewater:	N : P : K
	6 : 1 : 2

Chapter 5 presents a design equation developed from experience at municipal SR systems relating the K deficiency to the wastewater nitrogen and the needs of a particular crop. Industrial wastewaters are typically deficient in one or more of the critical nutrients. If SR or OF systems are planned for such effluents then supplemental fertilization with at least potassium is usually required.

TRACE ORGANICS

Public health impacts are the major concern with respect to land application of trace organics either through accumulation and persistance in the soil or movement to water supplies or the human food chain. The majority (88%) of the 128 items on the 1976 EPA list of priority pollutants are organic compounds. This is a strong indication of the growing concern with the fate of organics in the health environment and on the short and long term impacts on human health.

The removal of trace organics in land treatment results from a combination of volatilization, photodecomposition, adsorption and biological decomposition. The particular removal pathways that will dominate will depend on the type of compound involved (ie: volatile or stable) and on the degree of contact provided with the soil system.

Slow rate systems have been shown to be very effective in removal of trace organics. Table 3-18 presents results from the Muskegon, MI system (6) and from a pilot scale system in Hanover, NH (50).

Table 3.18: Trace Organic Removal in Slow Rate Systems (6, 50)

System Location Pollutant	.... Sampling Point/Mean Concentration, ppb				
	Wastewater	Aerated Lagoon	Storage Pond Effluent	After Sprinkling	Percolate at 5 ft +
Muskegon, MI					
Benzene	28	5	2	–	2
Chloroform	425	153	31	–	5
Trichloroethylene	52	18	6	–	6
Hanover, NH					
Chloroform	42	–	–	14	0.73
Toluene	57	–	–	24	0.02
Methylene chloride	8	–	–	2	0.04
Tetrachloroethylene	62	–	–	23	0.40

The overall removal at the Hanover system averaged 99.4% from all of the compounds tested. Approximately 65% of that removal occurred during the low pressure, large droplet, sprinkling operation. Removal of the less volatile organics has also been shown to be very effective. A study at the Muskegon, MI land treatment site (51) identified 56 organic compounds in the raw sewage. This is higher than would be expected for municipal wastewater but 60% of the flow in Muskegon comes from industrial sources. Only 17 of the compounds were present after aeration and detention in the storage ponds. Five of these were found in very low concentrations in the renovated water recovered from the site underdrains. Intake of these organics by plants grown on such sites is considered unlikely. The large molecular weight organics do not tend to pass the semipermeable membrane of plant roots. Surface adherence to the crop is possible. However, the production of vegetables to be eaten raw is not encouraged and a lag time is recommended prior to access for grazing animals.

Non Volatile Organics

There is evidence that the more stable organics such as pesticides and herbicides are adsorbed in the soil and eventually biodegraded, even at the very high hydraulic loading rates common to RI systems. The herbicide and pesticide removals observed at four recently studied land application systems were consistantly high (17). At three of the sites the concentrations of lindane in the application site groundwater was higher than background. However, in every case the concentration of lindane, 2,4-D, and 2,4,5-TP Silvex were all well below the limitations proposed for drinking waters. Endrin, methoxychlor and toxaphene were never found in the wastewaters at any site above the detection limits. Unless there were a spill from an industrial plant producing these materials it is unlikely that excessive concentrations will appear in municipal wastewaters. Such pesticides and herbicides are more likely to be used during the agricultural management of slow rate sites producing crops for the market. In these cases they would be used at rates that do not exceed normal agronomic practice.

Chlorinated Hydrocarbons

The fate of chlorinated hydrocarbons is a particular concern because of the carcinogenic risks in water supplies. Application of chlorinated hydrocarbons to a land treatment site can occur if the wastewater has an industrial component or as a direct result of chlorine disinfection prior to application to the land.

The transport of trace organics has been studied at the 40 year old RI system at Ft. Devens, MA (52). Twenty-five different organic compounds were measured in the primary effluent applied to the basins. Seventeen of these were then detected in the down gradient sampling wells in very low concentrations. For example, toluene was present in the primary effluent at 89 micrograms per liter but only 0.01 micrograms per liter in the sample well.

It can be concluded from the studies discussed above that

land treatment can be very effective in the removal of trace organics. It would appear that SR and RI will provide superior treatment for organics than OF and will provide greater environmental protection than direct stream discharge of municipal effluents. There should be no adverse impacts on the aquifers beneath land treatment sites from the trace organics in municipal wastewaters. In fact, chlorination prior to land application should be avoided to further enhance trace organics removal. No additional treatment for organics should be required for groundwaters recovered at the boundaries of SR and RI systems. One exception might be the recovery of percolate immediately beneath RI sites, if that recovered water is then intended for direct human consumption. It would be prudent in such a case to incorporate activated carbon treatment just prior to reuse as an added safety factor.

REFERENCES

1. Hossner, L. R., et al, Sewage Disposal on Agricultural Soils San Angelo, TX, EPA 600/2-78-131a, ORD, US EPA. (1978).
2. Koerner, E. L., Haws, D.A., Long-Term Effects of Land Application of Domestic Wastewater-Roswell, NM, EPA 600/2-79-047. (1979).
3. Culp, G.L., Hinrichs, D.J., A Review of the Operation of the Muskegon County Wastewater Management System, Muskegon County Board of Public Works, (Sept. 1978).
4. Benham-Blair, et al., Long-Term Effects of Land Application of Domestic Wastewater: Dickinson, ND-Slow Rate. U.S. EPA, ORD EPA 600/2-79-144. (1979).
5. Jenkins, T.F., Palazzo, A.J., Wastewater Treatment by a Slow Rate Land Treatment System; USA CRREL Report 81-14, (Aug. 1981).
6. U.S. EPA, Process Design Manual for Land Treatment of Municipal Wastewaters, EPA 625/1-81-013, (Oct. 1981).

7. Leach, L.E., Enfield, C.G., Harlin, C.C., Summary of Long-Term Rapid Infiltration System Studies, U.S. EPA, EPA 600/2-80-165, (July 1980).

8. Smith, R.G., Schroeder, E.D., Demonstration of the Overland Flow Process for the Treatment of Municipal Wastewater-Phase II Field Studies. Dept. Civil Engineering, Univ. of California, Davis. (August 1982).

9. Jewell, W.J., Limitations of Land Treatment of Wastes in the Vegetable Processing Industries, Cornell University (1978).

10. Smith, J.H., et al, "Treatment of Potato Processing Wastewater on Agricultural Land: Water and Organic Loading and the Fate of Applied Plant Nutrients in: Land as a Waste Management Alternative. Loehr, R.C., (ed.) Ann Arbor Science. (1977).

11. Perry, L.E., Reap, E.J., Gilliland, M., Pilot Scale Overland Flow Treatment of High Strength Snack Food Processing Wastewaters. Proceedings ASCE EED Conference, Atlanta, GA, p. 460-467, (July 8-10, 1981).

12. Wallace, A.T., et al, Rapid-Infiltration Disposal of Kraft Mill Effluent in: Proceedings 30th Industrial Waste Conference, p. 506-516, Purdue University, (May 6-8, 1975).

13. Crites, R.W., Land Treatment and Reuse of Food Processing Waste. Proceedings of the Industrial Wastes Symposia. WPCF Conference, St. Louis, MO (October 3-8, 1982).

14. U.S. Environmental Protection Agency. Pollution Abatement in the Fruit and Vegetable Industry. Environmental Research Information Center. EPA-625/3-77-007. (July 1977).

15. Witherow, J.L., Bledsoe, B.E., Algae Removal by the Overland Flow Process, Journal WPCF 55 (10):1256-1262 (1983).

16. Parsons, W.C., Spray Irrigation of Wastes from the Manufacture of Hardboard, Proceedings Purdue Industrial Waste Conference XXII, p. 602-607, (May 1967).

17. Reed, S.C., Health Aspects of Land Treatment, U.S. EPA CERI, U.S. GPO, 1979-657-093/7086, 43 p. (June 1979).

18. Crites, R.W., Wastewater and Sludge Use on Land, Environmental Science and Technology, in press, (1984).

19. Hays, B.D., Potential for Parasitic Disease Transmission with Land Application of Sewage Plant Effluents and Sludges, Journal IAWPR, 11:583-595. (1977).

20. Dean, R.B., The Sewage Farms of Paris, Proceedings Land Treatment Symposium, USA CRREL (1978).

21. Popp, L., Irrigation with Sewage from the Hygenic Point-of-View, International Commission on Irrigation and Drainage, 5th Regional Meeting, Cambridge, England. (1967).

22. Bell, R.G., Bole, J.B., Elimination of Fecal Coliform Bacteria from Soil Irrigated with Municipal Sewage Lagoon Effluent, Journal Environ. Qual. 7:193-196. (1978).

23. Deemer, D.O., Overland Flow, U.S. EPA Technology Transfer Seminar on Land Treatment, (June 1979).

24. Merrell, J.C. et al., The Santee Recreation Project, Santee, CA; Report WP-20-7, Fed. Water Pollut. Control Adm. (1967).

25. Vaughn, J.M., Landry, E.F., The Occurrence of Human Enteroviruses in a Long Island Groundwater Aquifer Recharged with Tertiary Wastewater Effluents, Proceedings, Land Treatment Symposium, USA, CRREL. (1978).

26. Lance, J.C., Gerba, C.P., Poliovirus Movement During High Rate Land Filtration of Sewage Water, Journal Environ. Qual., 9(1): 31-34, (1980).

27. Lennette, E.H., Problems Posed to Man by Viruses in Municipal Wastes, Virus Survival in Water and Wastewater Systems, Ed. Malina and Sagik, Univ. Tex., (1976).

28. Adams, A.P. et al, Bacterial Aerosols from Cooling Towers, Journal WPCF, 50(10):2362-2369. (1978).

29. Camann, D.E., et al, Evaluating the Microbiological Hazard of Wastewater Aerosols, Contract Report DAMD 17-75-C-5072, U.S. AMBRDL, Ft. Detrick, MD. (1978).

30. Katzenelson, E., Teltch, B., Dispersion of Enteric Bacteria by Spray Irrigation, Journal WPCF, 48(4):710-716. (1976).

31. Shuval, H.I., Teltch, B., Hygenic Aspects of the Dispersion of the Enteric Bacteria and Viruses by Sprinkled Irrigation of Wastewater, Water Reuse Symposium, AWWA. (1979).

32. U.S. EPA, Estimating Microorganism Densities in Aerosols from Spray Irrigation of Wastewater, U.S. EPA-600/9-82-003 (May 1982).

33. Middlebrooks, E.J., Industrial Pollution Control, Vol. 1 Agro. Industries, J. Wiley & Sons, NY, NY (1979).

34. Overcash, M.R., Pal, D., Design of Land Treatment Systems for Industrial Wastes, Ann Arbor Science, (1979).

35. Kincannon, C.B., Oily Waste Disposal by Soil Cultivation Process, U.S. EPA, EPA-R2-72-110. (1972).

36. U.S. EPA, Municipal Sludge Management: Environmental Factors. U.S. EPA, EPA-430/ 9-77-004. (Oct. 1977).

37. Uiga, A., Crites, R.W., Relative Health Risks of Activated Sludge Treatment and Slow Rate Land Treatment. Journal WPCF, 52(12): 2865-2874. (Dec. 1980).

38. Hinesly, T. D., et al, Effects on Corn by Applications of Heated Anaerobically Digested Sludge. Compost Science, 13(4):26-30, (1972).

39. Hinesly, T. D., et al, Soybean Yield Responses and Assimiliation of Zn and Cd from Sewage Sludge Amended Soil. Journal WPCF, 48:2137-2152. (1976).

40. Anderson, N., Notice Paper No. 15, Victoria, Australia, Legislative Assembly, (June 2, 1976).

41. Anthony, R.G., et al, Effects of Municipal Wastewater Irrigation on Selected Species of Animals, Proceedings Land Treatment Symposium, U.S.A. CRREL. (1978).

42. Pound, C.E., Crites, R.W., Olson, J.V., Long-Term Effects of Land Application of Domestic Wastewater-Hollister, CA, Rapid Infiltration Site. U.S. EPA. EPA-600/2-78-084. (April 1978).

43. Parker, C.F., L.E. Sommers, Mineralization of Nitrogen in Sewage Sludges, Journal Env. Quality, 12(1):150-156 (1983).

44. U.S. EPA, Process Design Manual for Utilization of Municipal Wastewater Sludge on Land, EPA 625/1-83-016, (Oct. 1983).

45. Crites, R.W., et al, Nutrient Control, WPCF MOP #FD-7, Chapter 9-Land Treatment. (1983).

46. Reed, S.C., ed. Wastewater Management by Disposal on the Land. USA CRREL Special Report 171, (May 1972).

47. Cornell University Training Course, Land Treatment (1976).

48. U.S. DOI, Drainage Manual, U.S. Dept. of Interior, GPO Stock No. 024-003-00177-1, (1978).

49. Ayers, R.S., Quality of Water for Irrigation, Journal, IRD ASCE. 103 (IRZ): 135-154, (1977).

50. Jenkins, T.F., Palazzo, A.J., Wastewater Treatment by a Prototype Slow Rate Land Treatment System. U.S. Army Corps of Engineers. CRREL Report 81-14. Hanover, NH (Aug. 1981).

51. U.S. EPA, Preliminary Survey of Toxic Pollutants at the Muskegon Wastewater Management System, ORD U.S. EPA. (1977).

52. Bedient, P.B., et al., Groundwater Transport from Wastewater Infiltration. Journal EED, ASCE 109(2):485, (April 1983).

4

Hydraulics of Soil Systems

The hydraulic capacity of the soil to accept and transmit water is crucial to the design of rapid infiltration (RI) systems and important in the design of most slow rate (SR) systems. The important hydraulic factors are infiltration, vertical permeability (percolation), horizontal permeability, groundwater mounding, and the relationship between predicted capacity and actual operating rates.

SOIL PROPERTIES

The hydraulics of soil systems are controlled by the physical and chemical properties of soil. Important physical properties include texture, structure, and soil depth. Chemical characteristics that can be important include pH, organic matter and exchangeable sodium percentage. Information on these soil properties and on soil permeability can be obtained from the Soil Conservation Service (SCS) and their detailed soil surveys.

Soil surveys will normally provide soil maps delineating the apparent boundaries of soil series with their surface texture. A written description of each soil series provides limited information on chemical properties, engineering applications, interpretive and management information, slopes, drainage, erosion potentials, and general suitability for most kinds of crops grown in the particular area. Additional information on soil characteristics and information regarding the availability of soil surveys can be obtained directly from the SCS. The SCS serves as the coordinating agency for the National Cooperative Soil Survey, and as such, cooperates with other government agencies, universities, and agricultural extension services in obtaining and distributing soil survey information.

Soil Physical Properties

The physical properties of texture and structure are important because of their effect on hydraulic properties. Soil textural classes are defined on the basis of the relative percentage of the three classes of particle size--sand, silt, and clay. Sand particles range in size from 2.0 mm to 0.05 mm; silt particles range from 0.05 mm to 0.002 mm; and particles smaller than 0.002 mm are clay. From the particle size distribution, the textural class can be determined using the textural triangle shown in Figure 4-1.

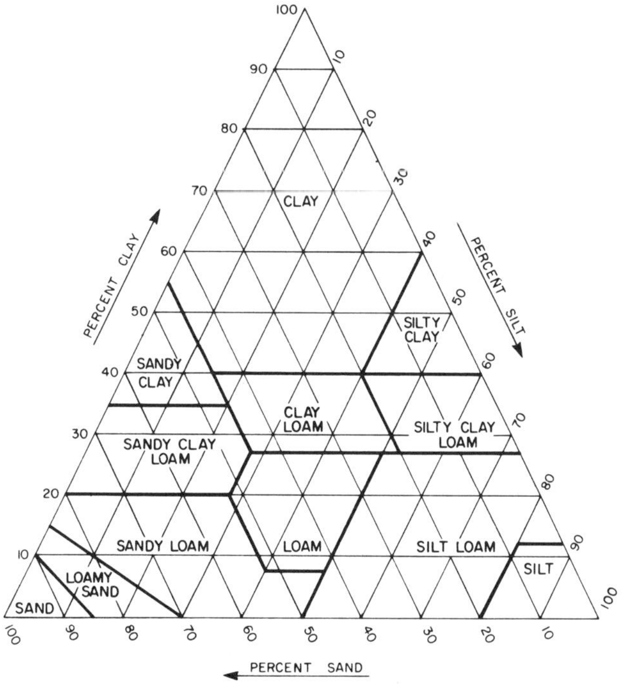

Figure 4-1 Soil Conservation Service (SCS) soil textural classes. After (14).

Fine-textured soils do not drain well and retain large percentages of water for long periods of time. As a result, crop management is more difficult than with more freely drained soils such as loamy soils. Fine-textured soils are generally best suited to overland flow systems. Medium-textured soils exhibit the best balance for wastewater renovation and drainage. Loam (medium texture) soils are generally best suited for slow rate systems.

Coarse-textured soils (sandy soils) can accept large quantities of water and do not retain moisture very long. This feature is important for crops that cannot withstand prolonged submergence or saturated root zones. Soil structure refers to the aggregation of individual soil particles. If these aggregates resist disintegration when the soil is wetted or tilled, it is well structured. The large pores in well-structured soils conduct water and air, making well-structured soils desirable for infiltration.

Adequate soil depth is needed for retention of wastewater constituents on soil particles, for plant root development, and for bacterial action. Retention of wastewater constituents, as explained in Chapter 3, is a function of residence time of wastewater in the soil. Residence time depends on the application rate and the soil permeability.

The type of land treatment process being considered will determine the minimum acceptable soil depth. For SR, the soil depth can be 2 to 5 ft, depending on the soil texture and crop type. For example, soil depths of 1 to 2 ft can support grass or turf, whereas deep rooted crops do better on soil depths of 4 to 5 ft.

The soil depth for RI should be at least 5 ft and preferably 5 to 10 ft. Overland flow systems require sufficient soil depth to form slopes that are uniform and to maintain a vegetative cover. A finished slope should have a minimum of 6 to 12 in. of soil depth.

Soil Chemical Properties

Soil chemical properties affect plant growth, wastewater

renovation and can affect hydraulic conductivity. Soil pH affects plant growth, bacterial growth and retention of elements such as phosphorus in the soil. Organic matter can improve soil structure and thereby improve the hydraulic conductivity.

Sodium can reduce the hydraulic conductivity of soil by dispersing clay particles and destroying the structure that allows water movement. Soils containing excessive exchangeable sodium are termed "sodic" or "alkali". A soil is considered sodic if the percentage of the total cation exchange capacity (CEC) occupied by sodium, the exchangeable sodium percentage (ESP), exceeds 15%. Fine-textured soils may be affected at an ESP above 10%, but coarse-textured soil may not be damaged until the ESP reaches about 20%. See Chapter 3 for additional discussion of sodium.

WATER MOVEMENT IN SOIL

Infiltration Rate

The rate at which water enters the soil surface, measured in in/h, is the infiltration rate. The infiltration rate is usually higher at the beginning of water application than it is several hours later. Infiltration rates are related to the extent of large, interconnected pore spaces in the soil. Coarse textured soils with many large pores have higher infiltration rates than fine textured-soils or soils in which the pore space is reduced in size by compaction or a breakdown of soil aggregates.

For a given soil, initial infiltration rates may vary considerable, depending on the initial soil moisture level. Dry soil has a higher initial rate than wet soil because there is more empty pore space for water to enter. The short term decrease in infiltration rate is primarily due to the change in soil structure and the filling of large pores as clay particles absorb water and swell. Thus, adequate time must be allowed when running field tests to achieve a steady intake rate.

Infiltration rates are affected by the ionic composition of the soil-water, the type of vegetation, and tillage of the soil surface. Factors that have a tendency to reduce infiltration rates include clogging by suspended solids in wastewater, classification of fine soil particles, clogging due to biological growths, gases produced by soil microbes, swelling of soil colloids, and air entrapped during a wetting event (2)(3). These influences are all likely to be experienced when a site is developed into a land treatment system. The net result is to restrict the hydraulic loadings of land treatment systems to values substantially less than those predicted from the steady state intake rates, requiring reliance on field-developed correlations between clean water infiltration rates and satisfactory operating rates for full-scale systems. It should be recognized that good soil management practices can maintain or even increase operating rates, whereas poor practices can lead to substantial decreases.

Intake

The rate at which water in a furrow enters the soil is referred to as the intake rate (4). Irrigation texts have used the term "basic intake rate" as synonymous with infiltration rate (5). In furrow irrigation the intake rate is influenced by the furrow size and shape. Therefore, when the configuration of the soil surface influences the rate of water entry, the term intake rate should be used rather than the term infiltration rate (which refers to a relatively level surface covered with water).

Permeability

The permeability or hydraulic conductivity (used interchangeably in this book) is the velocity of flow caused by a unit gradient. Permeability is not influenced by the gradient, and this is an important difference between infiltration and permeability.

Vertical permeability is also known as percolation. Lateral

flow is a function of the gradient and the horizontal permeability (which is generally different than the percolation rate). Permeability is affected mostly by the soil physical properties. Changes in water temperature can affect permeability slightly (4).

Transmissivity

Transmissivity of an aquifer is the product of the permeability (K) and the aquifer thickness. It is the rate at which water is transmitted through a unit width of aquifer under a unit hydraulic gradient.

Specific Yield

The term specific yield is the volume of water released from a known volume of saturated soil under the force of gravity and inherent soil tension (1). The specific yield is also referred to as the storage coefficient and the drainable voids. The primary use of specific yield is in aquifer calculations such as drainage and mound height analyses.

For relatively coarse-grained soils and deep water tables, it is usually satisfactory to consider the specific yield a constant value. As computations are not extremely sensitive to small changes in the value of specific yield, it is usually satisfactory to estimate it from knowledge of other soil properties, either physical as in Figure 4-2 (6), or hydraulic as in Figure 4-3 (1). To clarify Figure 4-2, specific retention is equal to the porosity minus the specific yield.

Water-Holding Capacity

Soil water can be classified as hygroscopic, capillary, and gravitational. Hygroscopic water is on the surface of soil particles and is not removed by gravity or by capillary forces. Capillary water is the water held in soil pores against gravity. Gravitational water is the water that will drain by gravity if favorable drainage is provided (4).

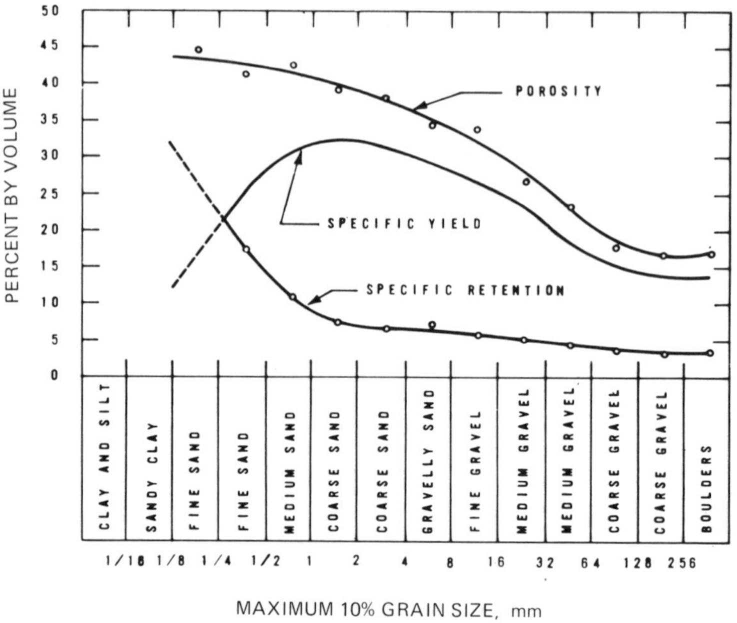

Figure 4-2 Porosity, specific yield, specific retention vs soil grain size, After (11).

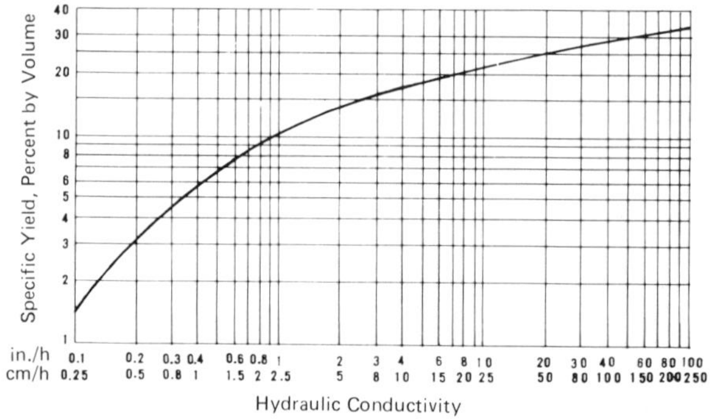

Figure 4-3 Specific yield vs hydraulic conductivity, After (11).

Soil water can also be classified according to its availability to plant root systems. As illustrated in Figure 4-4, the maximum available water occurs at saturation (point 1), when all the pore space is filled with water. When the soil water drops to point 3, only hygroscopic water is left, which is unavailable to plants.

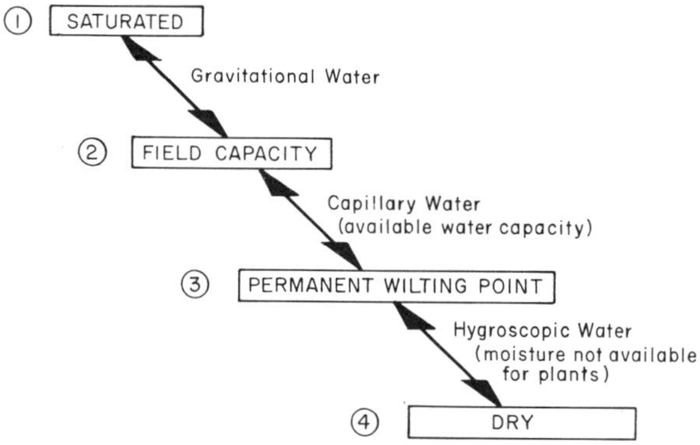

Figure 4-4 Soil moisture characteristics.

<u>Field Capacity</u>: When gravitational water has been removed, the moisture content of the soil is called the field capacity. In practice the field capacity is measured 2 days after water application and can range from 3% moisture for fine sand to 40% for clay. The range of moisture percentages for field capacity for various soil types is presented in Table 4-1. Relationships of field conditions to soil moisture content are presented in Table 4-2.

<u>Permanent Wilting Point</u>: The soil moisture content at which plants will wilt from lack of water is known as the permanent wilting point. The available moisture content is generally defined as the difference between the field capacity and the permanent wilting point. This represents the moisture that can be stored in the soil for subsequent use by plants. For SR systems with poorly drained soils, this stored moisture is important to design loadings.

As an approximation the permanent wilting percentage can be obtained by dividing the field capacity by 2. For soils with high silt content, divide the field capacity by 2.4 to obtain permanent wilting percentage.

Table 4.1: Range of Available Soil Moisture for Different Soil Types (2)

Soil type	Moisture Percentage: Field Capacity	Moisture Percentage: Permanent Wilting Point	Depth of Available Water per Unit Depth of Soil (in./ft)
Fine sand	3-5	1-3	0.3-0.5
Sandy loam	5-15	3-8	0.5-1.3
Silt loam	12-18	6-10	0.7-1.6
Clay loam	15-30	7-16	1.2-2.2
Clay	25-40	12-20	2.0-3.5

Table 4.2: Field Estimating of Soil Moisture Content[a]

Fine Texture	Medium Texture	Moderately Coarse Texture	Coarse Texture
No free water after squeezing, wet, outline on hand	Same as fine texture	Same as fine texture	Same as fine texture
0.0	0.0	0.0	0.0
Easily ribbons out between fingers, has slick feeling	Forms a very pliable ball, sticks readily if high in clay	Forms weak ball, breaks easily, will not stick	Sticks together slightly may form a very weak ball under pressure
0.0-0.6	0.0-0.5	0.0-0.4	0.0-0.2
Forms a ball, ribbons out between thumb and forefinger	Forms a ball, sometimes sticks slightly with pressure	Tends to ball under pressure, but will not hold together	Appears dry, will not form a ball when squeezed
0.6-1.2	0.5-1.0	0.4-0.8	0.2-0.5
Somewhat pliable, will form a ball when squeezed	Somewhat crumbly, but holds together from pressure	Appears dry, will not form a ball	Appears dry, will not form a ball
1.2-1.9	1.0-1.5	0.8-1.2	0.5-0.8
Hard, baked, cracked	Powdery, dry, sometimes slightly crusted but easily broken down into powdery condition	Dry, loose, flows through fingers	Dry, loose, single-grained flows through fingers
1.9-2.5	1.5-2.0	1.2-1.5	0.8-1.0

[a]The numerical values are the amount of water (in) that would be needed to bring the top foot of soil to field capacity.

SATURATED HYDRAULIC CONDUCTIVITY

In general, water moves through soils or porous media in accordance with Darcy's equation:

$$q = \frac{Q}{A} = K\frac{dh}{dl} \qquad (4\text{-}1)$$

where q = flux of water, the flow, Q per unit cross sectional area, A, in/h

K = hydraulic conductivity (permeability), in/h

dh/dl = hydraulic gradient, ft/ft

The total head (H) can be assumed to be the sum of the soil-water pressure head (h), and the head due to gravity (Z), or $H = h + Z$. The hydraulic gradient is the change in total head (dH) over the path length (dl).

The hydraulic conductivity is defined as the proportionality constant, K. The conductivity (K) is not a true constant but a rapidly changing function of water content. Even under conditions of constant water content, such as saturation, K may vary over time due to increased swelling of clay particles, change in pore size distribution due to classification of particles, and change in the chemical nature of soil-water. However, for most purposes, saturated conductivity (K) can be considered constant for a given soil. The K value for flow in the vertical direction will not necessarily be equal to K in the horizontal direction. This condition is known as anisotropic. It is especially apparent in layered soils and those with large structural units. An illustration of anisotropic conditions is shown in Table 4-3.

Table 4.3: Measured Ratios of Horizontal to Vertical Conductivity (8)(9)

Site	Horizontal Conductivity (K_h, ft/d)	K_h/K_v	Remarks
1	138	2.0	Silty
2	246	2.0	–
3	184	4.4	–
4	328	7.0	Gravelly
5	236	20.0	Near terminal moraine
6	236	10.0	Irregular succession of sand and gravel layers (from K measurements in field)
6	282	16.0	(From analysis of recharge flow system)

The value of K depends on the size and number of pores in the soil or aquifer material. Orders of magnitudes for vertical conductivity (K_v) values in ft/day for typical soils are (10):

Clay soils (surface)	0.03-0.06
Deep clay beds	$3x10^{-8}$ - 0.03
Clay, sand, and gravel mixes (till)	0.003 - 0.3
Loam soils (surface)	0.3 - 3.0
Fine sand	3 - 16
Medium sand	16 - 66
Coarse sand	66 - 300
Sand and gravel mixes	16 - 330
Gravel	330 - 3300

Example 4-1: Subsurface Flow

Conditions: The soil on the hillside (Figure 4-5) is a loam with a hydraulic conductivity as shown in the sketch.

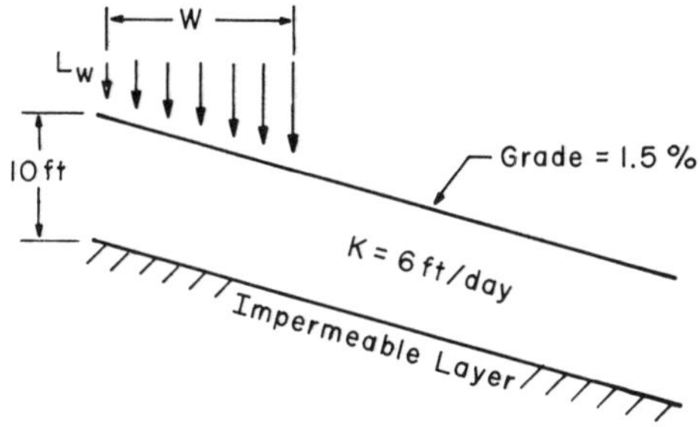

Figure 4-5 Subsurface flow for Example 4-1.

Solution: Calculate the transmissivity and the unit flow rate under saturated conditions.

1. Since flow is essentially in the horizontal direction, the grade of the impermeable layer determines the hydraulic gradient.

$$T = K D \qquad (4\text{-}2)$$

where T = Transmissivity, ft^2/day

K = Hydraulic conductivity, ft/day

D = Depth of aquifer, ft

$$T = (6\ ft/day)\ (10\ ft)$$
$$= 60\ ft^2/day$$

Converting to gpd, multiply by 7.48

$$T = 448.8\ gpd/ft$$

The flow under saturated conditions is

$$q = Ti$$

where q = gpd/ft

T = gpd/ft

i = gradient, ft/ ft

$$q = 448.8\ (0.015)$$
$$6.73\ gpd/ft$$

2. The combination of loading rate, L_w, and width of application area, W, cannot exceed the saturated flow q = 6.73 gpd/ft. For a loading rate of 0. 1 ft/d, calculate the width, W.

$$L_wW = \frac{6.73}{7.48} = 0.9\ ft^2/d$$

$$W = \frac{0.9}{L_w}$$

$$W = \frac{0.9}{0.1}$$

$$W = 9\ ft$$

UNSATURATED HYDRAULIC CONDUCTIVITY

Darcy's law for velocity of flow in saturated soils applies also to unsaturated soils. As the moisture content decreases, however, the cross-sectional area through which the flow occurs also decreases and the conductivity is reduced.

The conductivity of soil varies dramatically as water content is reduced below saturation. As an air phase is now present, the flow channel is changed radically and now consists of an irregular solid boundary and the air-water interface. The flow path becomes more and more tortuous with decreasing water content as the larger pores empty and flow becomes confined to the smaller pores. Compounding the effect of decreasing cross-sectional area for flow is the effect of added friction as the flow takes place closer and closer to solid particle surfaces. The conductivity of sandy soils, although much higher at saturation than loam soils, decreases more rapidly as the soil becomes less saturated. In most cases, the conductivities of sandy soils eventually become lower than finer soils. This relationship explains why a wetting front moves more slowly in sandy soils than medium or fine textured soils after irrigation has stopped and why there is little horizontal spreading of moisture in sandy soils after irrigation.

PERCOLATION CAPACITY

The percolation capacity of SR and RI systems is a critical parameter in planning, design, and operation. The capacity will vary within a given site and may change with time, season and different management. For planning purposes the infiltration capacity can be estimated from the vertical permeability rates assigned by the SCS (Figure 4-6).

Design Percolation Rate

To account for the needed drying time between applications, the variability of the actual soil permeability within a site, and the potential reduction with time, a small percentage of the vertical permeability is used as the design percolation rate. This small percentage ranges from 4 to 10% of the saturated vertical permeability as shown in Figure 4-6. The value used for clear water permeability should be for the most restrictive layer in the soil profile. Design rates based on field measurement (Chapter 7) may be calculated using different percentages (Chapters 10 and 12).

Example 4-2: Determining Design Percolation Rate

Conditions: Given a soil with a permeability (most restrictive layer) of 0.6 to 2 in/h (moderate permeability). Determine the design wastewater percolation rate.

Solution: Using the 0.6 in/h rate, enter Figure 4-6 and proceed vertically to the hatched area. For a conservative value, proceed horizontally to the left to a value of 17 ft/yr. To obtain the maximum value recommended for planning proceed vertically to the top of the hatched area (10% value) and pick off the design percolation rate of 42 ft/yr. If the planned application season is less than 365 days, the percolation rate should be reduced to coincide with the planned application period.

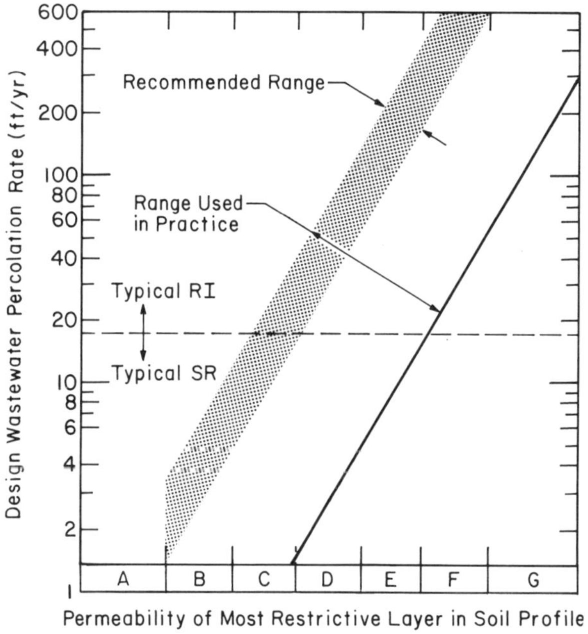

Figure 4-6 Design percolation rate vs SCS soil permeability for SR and RI. The zones A through G refer to clearwater permeability of most restrictive layer in the soil profile (K_v = in/h): A = very slow, < 0.06; B = slow, 0.06 to 0.20; C = moderately slow; 0.20 to 0.60; D = moderate, 0.60 to 2.0; E = moderately rapid, 2.0 to 6.0; F = rapid, 6.0 to 20, G = very rapid, > 20, After (11).

Calculation of Vertical Permeability

The rate at which water percolates through soil depends on the average saturated permeability (K) of the profile. If the soil is uniform, K is assumed to be constant with depth. Any differences in measured values of K are then due to normal variations in the measurement technique. Thus, average K may be computed as the arithmetic mean of n samples:

$$K_{am} = \frac{K_1 + K_2 + K_3 + \ldots + K_n}{n} \qquad (4\text{-}3)$$

where K_{am} = arithmetic mean vertical conductivity

Many soil profiles approximate a layered series of uniform soils with distinctly different K values, generally decreasing with depth. For such cases, it can be shown that average K is represented by the harmonic mean of the K values from each layer (12):

$$K_{hm} = \frac{D}{\frac{d_1}{K_1} + \frac{d_2}{K_2} + \ldots \frac{d_n}{K_n}} \qquad (4\text{-}4)$$

where D = soil profile depth

d_n = depth of nth layer

K_{hm} = harmonic mean conductivity

If a bias or preference for a certain K value is not indicated by statistical analysis of field test results, a random distribution of K for a certain layer or soil region must be assumed. In such cases, it has been shown that the geometric mean provides the best and most conservative estimate of the true K (12)(13)(14):

$$K_{gm} = (K_1 \cdot K_2 \cdot K_3 \cdot \ldots K_n)^{1/n} \qquad (4\text{-}5)$$

where K_{gm} = geometric mean conductivity

Example 4-3: Geometric Mean Calculation of Permeability

Conditions: Consider a soil profile with vertical permeabilities of K_1 = 2 in/h, K_2 = 0.6 in/h, and K_3 = 4 in/h.

Solution: Calculate the geometric mean conductivity

$$K_{gm} = \left[(2)(0.6)(4)\right]^{1/3}$$

$$K_{gm} = 1.69 \text{ in/h}$$

Profile Drainage

For SR and RI systems the soil profile must drain between applications to allow the soil to reaerate. The time required for profile drainage is important to system design and varies with the soil texture and the presence of restrictions (such as fragipans, clay pans, and hardpans). In sandy soils without vertical restrictions, the profile can drain in 1 to 2 d. In clayey soils drainage may take 5 d or more. The drying period between applications also depends on the evaporation rate (Chapter 5).

GROUNDWATER MOUNDING

If water that infiltrates the soil and percolates vertically through the zone of aeration encounters a water table or an impermeable (or less permeable) layer, a groundwater "mound" will begin to grow (See Figure 4-7).

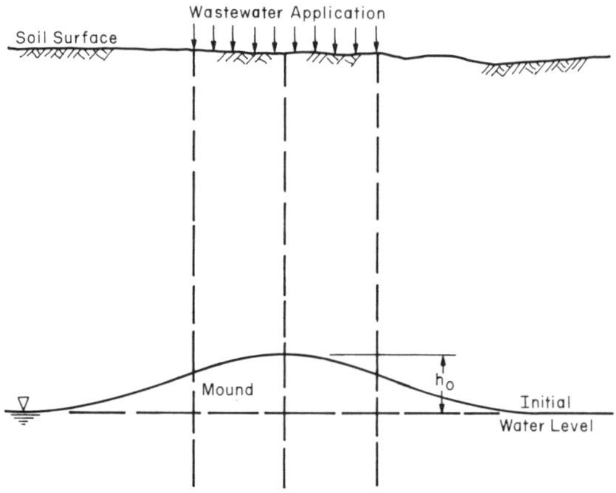

Figure 4-7 Schematic of groundwater mound.

If the mound height continues to grow, it may eventually encroach on the zone of aeration to the point where renovation capacity is affected. Further growth may result in intersection of the mound with the soil surface, which will reduce infiltration rates. This problem can usually be identified and analyzed before the system is designed and built if the proper geologic and hydrologic information is available for analysis.

Prediction of Mounding

Groundwater mounding can be estimated by applying heat-flow theory and the Dupuit-Forchheimer assumptions (13). These assumptions are as follows:

1. Flow within groundwater occurs along horizontal flow lines whose velocity is independent of depth.
2. The velocity along these horizontal streamlines is proportional to the slope of the free water surface.

Using these assumptions, heat-flow theory has been successfully compared to actual groundwater depths at several existing RI sites.

To compute the height at the center of the groundwater mound, one must calculate the values of $W/\sqrt{4\alpha t}$ and Rt.

where W = width of the recharge basin, ft

a = aquifer constant = $\frac{KD}{V}$, ft^2/d

where K = aquifer (horizontal) hydraulic conductivity, ft/d

D = saturated thickness of the aquifer, ft

V = specific yield or fillable pore space of the soil, ft^3/ft^3 (Figures 4-2 and 4-3)

t = length of wastewater application, d

R = I/V, ft/d, rate of rise if no lateral flow occurred

where I = application rate, ft/d

Once the value of $W/\sqrt{4\alpha t}$ is obtained, one can use dimensionless plots of $W\sqrt{4\alpha t}$ versus h_o/Rt, provided as Figures 4-8 (for square recharge areas) and 4-9 (for rectangular recharge areas), to obtain the value of h_o/Rt, where h_o is the rise at the center of the mound. Using the calculated value of Rt, one can solve for h_o.

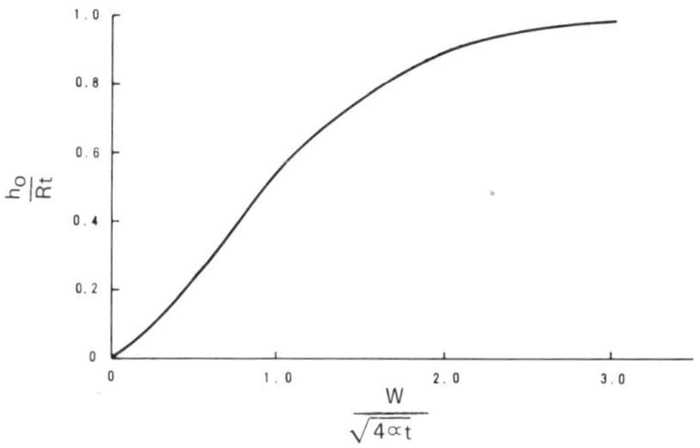

Figure 4-8 Mounding curve for center of a square recharge area, After (11).

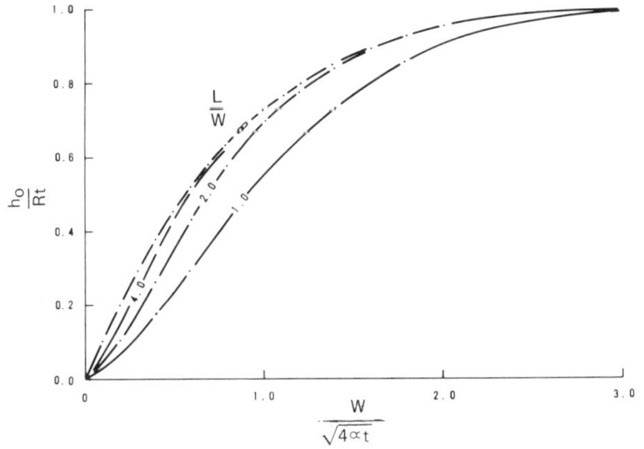

Figure 4-9 Mounding curve for center of a rectangular recharge area, with different ratios of length (L) to width (W), After (11).

<u>Example 4-4: Mound Height Analyses</u>

<u>Conditions</u>: Consider a situation where an RI system is pro-

posed with square infiltration basins. The saturated thickness, D, of the aquifer is 50 ft. The horizontal hydraulic conductivity, K, measured by the auger hole method (see Chapter 7) is 8 ft/d.

Solution: Using Figure 4-3 with this K value, the specific yield is found to be 17.5%. The basins are to be 100 ft wide and square; the application rate, I is 1 ft/d; and the application period, t, is 2 days.

1. First calculate the aquifer constant:

$$\alpha = \frac{KD}{V} = \frac{8(50)}{0.175} = 2{,}286\ \text{ft}^2/\text{d}$$

2. Next calculate the rate of rise, R.

$$R = \frac{I}{V} = \frac{1}{0.175} = 5.7\ \text{ft/d}$$

3. Then calculate the factor $W/\sqrt{4\alpha t}$

$$\frac{W}{\sqrt{4\alpha t}} = \frac{100}{\sqrt{4\ (2286)(2)}} = 0.74$$

Enter Figure 4-8 with the value of 0.74 on the abscissa and the resultant value of h_o/Rt equals 0.37.

4. Finally, calculate h_o, the height of the groundwater mound:

$$h_o = 0.37\ Rt = 0.37\ (5.7)(2) = 4.2\ \text{ft}$$

5. The initial depth to groundwater is 15 ft and the calculated mound height of 4.2 ft would bring the groundwater to within 10.8 ft of the ground surface. In this situation there would not be a need for engineered drainage. If the calculations should indicate that the groundwater table will rise to within 3 to 5 ft of the basin bottom, additional drainage will be needed.

Figures 4-10 (for square recharge areas) and 4-11 (for recharge areas that are twice as long as they are wide) can be used to estimate the depth to the mound at various distances from the center of the recharge basin. Again, the values of $W/\sqrt{4\alpha t}$ and Rt must be determined first. Then, for a given value of x/W, where x equals the horizontal distance from the center of the recharge basin, one can obtain the value of h_o/Rt from the correct plot. Multiplying this number by the calculated value of Rt results in the rise of the mound, H_o, at a distance x from the center of the recharge site. The depth to the mound from the soil surface is then the difference between the distance to the groundwater before recharge and the rise due to the mound.

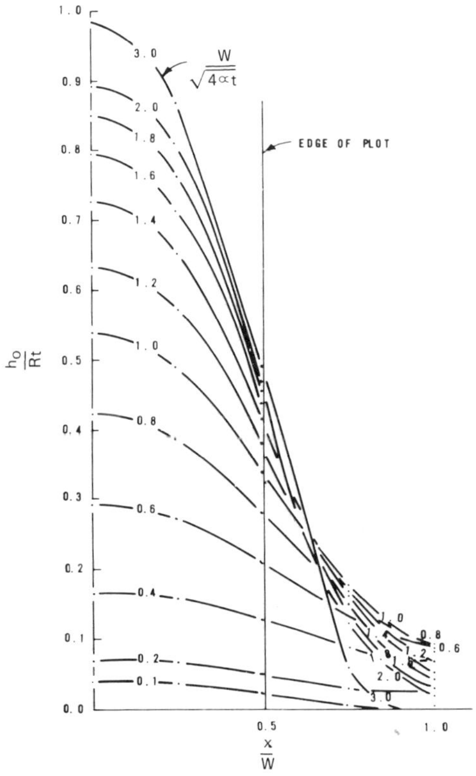

Figure 4-10 Rise and horizontal spread of a mound below a square recharge area, After (11).

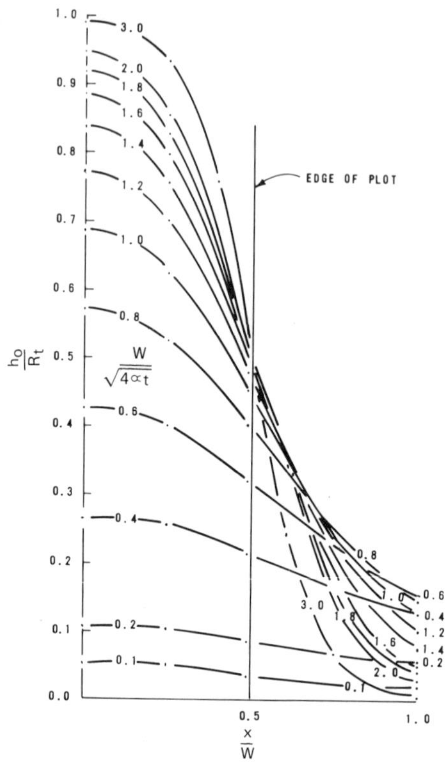

Figure 4-11 Rise and horizontal spread of mounds below a rectangular recharge area when L = 2W, (After 11).

To evaluate mounding beneath adjacent basins, Figures 4-10 and 4-11 should be used to plot groundwater table mounds as functions of distance from the center of the plot and time elapsed since initiation of wastewater application. Then, critical mounding times should be determined, such as when adjacent or relatively close basins are being flooded, and the mounding curves of each basin at these times should be superimposed. At sites where drainage is critical because of severe land limitations or extremely high groundwater tables, the engineer should use the approach described in reference (14) to evaluate mounding.

UNDERDRAIN SPACING

Generally, underdrains are spaced 50 ft or more apart. Depths of drains vary from 3 to 8 ft for SR systems and 8 to 15 ft for RI systems. In soils with high lateral permeability, the underdrains may be as much as 500 ft apart. The closer the drain spacing is, the more control there will be over depth of the groundwater table. The cost of drains increases with decreasing drain spacing, so the economics of using more drains must be weighed against finding a site with deeper groundwater, or less vertical restriction to percolation, or using a lower application rate.

One method of determining drain spacing is the Hooghhoudt method. The parameters used in the method are shown in Figure 4-12. The assumptions used in this method are (17):

1. The soil is homogeneous with a lateral permeability, K.
2. The drains are evenly spaced a distance S apart.
3. The hydraulic gradient at any point is equal to the slope of the water table above that point.
4. Darcy's Law is valid.
5. An impermeable layer underlies the drain at a depth d.
6. The rate of replenishment (wastewater application plus natural precipitation) is $L_W + P$.

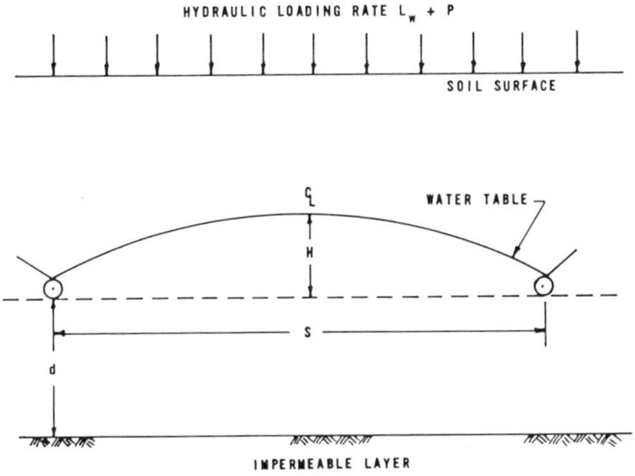

Figure 4-12 Parameters used in drain design (17).

To determine drain placement, the following equation is useful (17):

$$S = \left[\frac{4KH}{L_w + P}(2d + H)\right]^{0.5} \tag{4-6}$$

where S = drain space, ft

K = horizontal hydraulic conductivity of the soil, ft/d

H = height of the ground water mound above the drains, ft

L_w = annual wastewater loading rate, expressed as a daily rate, ft/d

P = average annual precipitation rate, expressed as a daily rate, ft/d

d = distance from drains to underlying impermeable layer, ft.

Example 4-5: Underdrain Spacing

Conditions: RI system to be loaded at 120 ft/yr or 0.33 ft/d. K = 20 ft/d, H = 3 ft, d = 2 ft, average precipitation is 0.02 ft/d.

Solution:

1. Calculate the hydraulic loading

$$L_w + P = 0.33 + 0.02$$
$$= 0.35 \text{ ft/d}$$

2. Calculate S

$$S = \left[\frac{4KH}{L_w + P}(2d + H)\right]^{0.5}$$

$$S = \left[\frac{(4)(20)(3)}{0.35}(2)(2) + 3\right]^{0.5}$$

$$S = \left[685.7\ (4 + 3)\right]^{0.5}$$

S = 69 ft.

REFERENCES

1. Drainage Manual, U.S. Department of the Interior, Bureau of Reclamation, 1st edition (1978).

2. Jarrett, A.R. and Fritton, D.D., Effect of Entrapped Soil Air on Infiltration. Transactions American Society of Agricultural Engineers. 21:901-906 (1978).

3. Parr, J.F. and Bertrand, A.R., Water Infiltation into Soils. In: Advances in Agronomy, Norman, A.G. (Ed.) New York; Academic Press, pp. 311-363 (1960).

4. Hansen, V.E., Israelson, O.W. and Stringham, G.E., Irrigation Principles and Practices, Fourth Edition New York; John Wiley & Sons (1980).

5. Pair, C.H., et al. Sprinkler Irrigation, Fourth Edition, Silver Spring, Maryland; Sprinkler Irrigation Association (1975).

6. Todd, D.K., Groundwater. In: Handbook of Applied Hydrology. Chow, V.T. (Ed.) New York: McGraw-Hill Book Co. (1964).

7. Booher, L.J. Surface Irrigation. FAO Agricultural Development Paper No. 95. Food and Agricultural Organization of the United Nations, Rome (1974).

8. Weeks, E.P. Determining the Ratio of Horizontal to Vertical Permeability by Aquifer-Test Analysis. Water Resources Research 5: 196-214 (1969).

9. Bouwer, H. Groundwater Recharge Design for Renovating Wastewater.

Journal Sanitary Engineering Division, ASCE 96 (SA-1): 59-73 (1970).

10. Bouwer, H., Groundwater Hydrology, New York: McGraw-Hill Book Co. (1978).

11. Process Design Manual for Land Treatment of Municipal Wastewater, U.S. Environmental Protection Agency, EPA 625/1-81-013 (October 1981).

12. Bouwer, H. Planning and Interpreting Soil Permeability Measurements. Journal Irrigation and Drainage Div. ASCE 28:391-402 (1969).

13. Rogowski, A.S., Watershed Physics: Soil Variability Criteria. Water Resources Research 8:1015-1023 (1972).

14. Nielson, D.R., Biggar, J.W. and Erb, K.T. Spatial Variability of Field-Measured Soil-Water Properties. Hilgardia. 42:215-259 (1973).

15. Bianchi, W.C. and Muckel, C., Ground-Water Recharge Hydrology. U.S. Department of Agriculture, Agricultural Research Service. ARS 41161 (December 1970).

16. Hantush, M.S., Growth and Decay of Groundwater-Mounds in Response to Uniform Percolation. Water Resources Research 3(1): 227-234 (1967).

17. Luthin, J.N., Drainage Engineering, Huntington, N.Y.: R.E. Krieger Pub. Co. (1973).

5

Vegetation As a Treatment Component

VEGETATION IN LAND TREATMENT

Vegetation plays different roles in each land treatment process. In SR the vegetation is essential and is generally used for nitrogen removal and, in some cases, for economic return. In OF vegetation is the support media for biological activity and is needed for erosion protection. The grass in OF systems also removes nutrients and slows the flow of wastewater so that suspended solids can be filtered and settled out of the flow stream.

Vegetation is not always part of RI systems. It can play a role in stabilization of the soil matrix and can maintain long-term infiltration rates, but does not appear to have a major impact on treatment performance for RI systems.

In this chapter the characteristics of crops that affect their use in land treatment--water use and tolerance, nutrient uptake, and toxicity concerns--are described. Guidance on crop selection for each land treatment process is provided. Crop management aspects of agronomic and forest crops are also described.

EVAPOTRANSPIRATION

Evapotranspiration (ET) is the combined loss of water from a given area by evaporation from the soil surface, snow, or intercepted precipitation, and by the transpiration and building of tissue by plants. Most water evaporated at plant surfaces is water transpired by the plant with only about 1% of the water taken up by plants actually consumed in the metabolic activity of the plant (1). The evapotranspiration rate is controlled by atmospheric demand and soil-water availability. If soil-water availability is sufficient, which it will be for land treatment, the potential rate of ET will be determined by solar radiation, air temperature, relative humidity and wind speed.

For land treatment systems the potential evapotranspiration is important in planning and design. Potential ET is the water lost from an extended surface of short green crop (reference crop) which fully shades the ground and is well supplied with water. Potential evapotranspiration cannot exceed free water evaporation under the same weather conditions (1).

Evaporation

Most data on evaporation have been obtained from evaporation pans, such as the U.S. Weather Bureau's Class A pan, which is 46.5 in. in diameter and 10 in. deep. Evaporation pans provide a measure of the combined effects of radiation, temperature, humidity and wind on evaporation from a specific open water surface. Pans store more heat than crops do, consequently, they cause evaporation measurements to be higher than the reference crop evapotranspiration (ETo). The pan coefficients, shown in Table 5-1, can be used to convert pan evaporation to reference crop evapotranspiration according to Equation 5-1.

$$ETo = K_{pan} \text{ Evap.} \qquad (5\text{-}1)$$

where Eto = reference crop evapotranspiration

K_{pan} = pan coefficient

Evap = pan evaporation

Table 5.1: Pan Coefficients for Class A Evaporation Pans Placed in a Reference Crop Area (2)

Wind, mph	.... Relative Humidity (%)		
	Low, <40	Medium, 40-70	High, >70
Light, 4.5	0.75	0.85	0.85
Moderate, 4.5	0.70	0.80	0.80
Strong, 11-18	0.65	0.70	0.75
Very strong, >18	0.55	0.60	0.65

Calculating Evapotranspiration

The crop ET used in planning is the seasonal total ET. Typical values of seasonal ET for different crops are presented in Table 5-2. In many states, estimates of seasonal ET for various crops can be obtained from local agricultural extension offices, the land grant university, agricultural research stations or the SCS. If crop ET data are not available, they can be calculated from the reference crop, ETo, using crop coefficients.

Table 5.2: Range of Seasonal Crop Evapotranspiration (2, 3, 4)

Crop	ET (inches)	Crop	ET (inches)
Alfalfa	24-74	Grass	18-45
Avocado	26-40	Oats	16-25
Barley	15-25	Potatoes	18-24
Beans	10-20	Rice	20-45
Clover	34-44	Sorghum	12-26
Corn	15-25	Soybeans	16-32
Cotton	22-37	Sugarbeets	18-33
Deciduous trees	21-41	Sugarcane	39-59
Grains (small)	12-18	Vegetables	10-20
Grapes	16-35	Wheat	16-28

Crop coefficients can change during the growing season depending on the crop planting date, rate of crop development, length of growing season and climatic conditions. For annual crops there are four different stages of crop development:

1. Initial growth stage (ground cover 10%)
2. Crop development stage (up to ground cover 80%)
3. Mid-season stage (effective full ground cover)

4. Late season stage (full maturity until harvest)

Each stage is characterized by a different crop coefficient. For the first two stages, the curves in Figure 5-1 can be used to estimate the crop coefficient. To use Figure 5-1 enter the plot with the reference crop ETo, proceed up to the recurrence interval between wastewater applications, and then pick off the value of kc at the left.

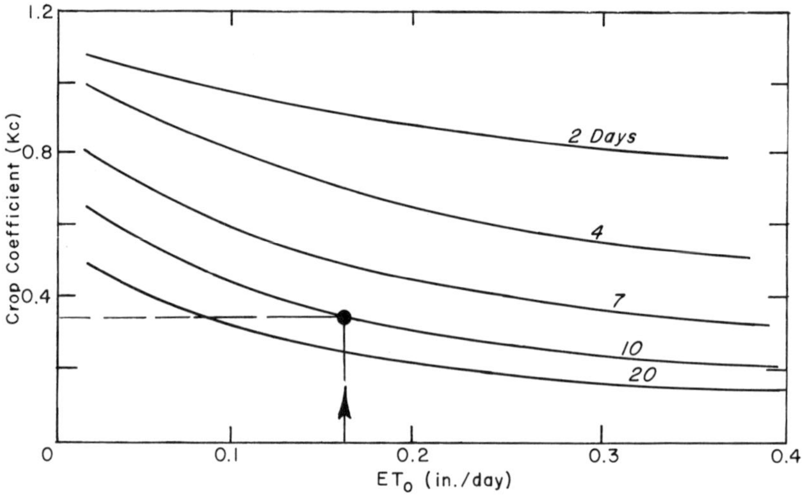

Figure 5-1 Average crop coefficient (Kc) values for initial, and crop development stages, the curves are for average recurrence interval of irrigation, or significant rain, After (2).

The third and fourth stages of crop growth produce the largest values of kc, as presented in Table 5-3. Ranges of lengths of each crop growth stage are presented in Table 5-4. To estimate the crop ET, determine the crop coefficient for each stage, multiply by the number of days in the period and multiply by the reference crop ET.

Table 5.3: Crop Coefficient, Kc, for Mid-Season and Late Season Conditions (2)

Crop	Crop Stage	Humid*	Dry**
Alfalfa***	1-4	0.85	0.95
Barley	3	1.05	1.15
	4	0.25	0.20
Clover	1-4	1.00	1.05
Corn	3	1.05	1.15
	4	0.55	0.60
Cotton	3	1.05	1.20
	4	0.65	0.65
Grain	3	1.05	1.15
	4	0.30	0.25
Grapes	3	0.80	0.90
	4	0.65	0.70
Oats	3	1.05	1.15
	4	0.25	0.20
Pasture grass	1-4	0.95	1.00
Rice	3	1.1	1.25
Sorghum	3	1.00	1.10
	4	0.50	0.55
Soybeans	3	1.00	1.10
	4	0.45	0.45
Sugarbeets	3	1.05	1.15
	4	0.90	1.00
Wheat	3	1.05	1.15
	4	0.25	0.20

*Humidity 70%, light wind 0-16 mph.
**Humidity 20%, light wind 0-16 mph.
***Peak factors are 1.05 for humid conditions and 1.15 for dry conditions.

Table 5.4: Length of Four Crop Growth Stages for Typical Annual Crops (days) (2)

	Growth Stage			
Crop	1	2	3	4
Barley	15	25-30	50-65	30-40
Corn	20-30	35-50	40-60	30-40
Cotton	30	50	55-60	45-55
Grain, small	20-25	30-35	60-65	40
Sorghum	20	30-35	40-45	30
Soybeans	20	30-35	60	25
Sugar beets	25-45	35-60	50-80	30-50

Example 5-1: Crop ET Calculation

Conditions: Estimate the growing season ET for corn (grain) planted in mid-May. Winds are light and the humidity is low (less than 20%). The reference crop ETo is 0.20 in/d during the 20 day initial development stage and the 35 day crop development stage.

The ETo increases to 0.25 in/d in the 40 day stage 3 and then declines to 0.20 in/d for the 30 d stage 4. The application frequency is 10 d in the first two stages.

Solution: A plot of the crop coefficient versus growing period is presented in Figure 5-2. The growing season ET is as follows:

Stage 1

Enter Figure 5-1 at ETo = 0.16 in/d and proceed to the recurrence interval curve for 10 d. Move horizontally to the left and pick off the value of kc = 0.35.

ET = (kc)(days)(ETo)

= (0.35)(20d)(0.20)in/d)

= 1.40 in.

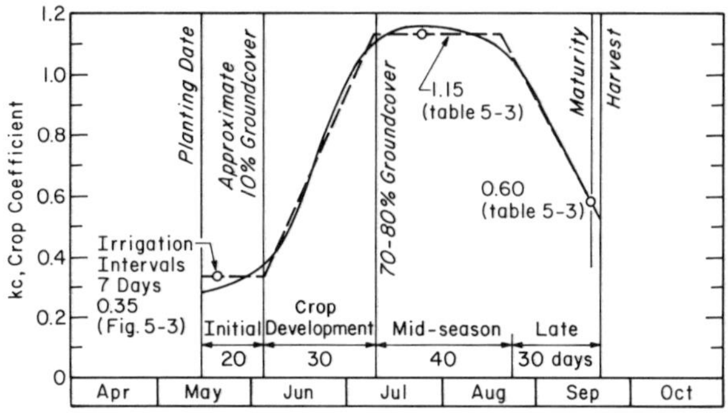

Figure 5-2 Sample crop coefficient curve for corn, After (2).

Stage 2

For the second stage the value of kc increases from 0.35 to 1.15 (stage 3 value from Table 5-3). As shown on Figure 5-2 the increase can be estimated using a straight line between the first and third stages:

$$kc = \frac{0.35 + 1.15}{2}$$

$$= 0.75$$

and

$$ET = (0.75)(35d)(0.20 \text{ in./d})$$
$$= 5.25 \text{ in.}$$

Stage 3

From Table 5-3 the value of kc for dry conditions is 1.15.

$$ET = (1.15)(40 \text{ d})(0.25 \text{ in./d})$$
$$= 11.50 \text{ in.}$$

Stage 4

From Table 5-3 the stage 4 value of kc is 0.6.

Using the graph in Figure 5-2, the average value of kc in stage 4 is:

$$kc = \frac{1.15 + 0.6}{2}$$
$$= 0.875$$

and

$$ET = (0.875)(30d)(0.2 \text{ in./d})$$
$$= 5.25 \text{ in.}$$

Total ET for this 125 d growing season is:

$$1.40 + 5.25 + 11.50 + 5.25 = 23.4 \text{ in.}$$

Potential Evapotranspiration

In humid regions estimates of potential evapotranspiration are usually sufficient for crop water use for perennial full-cover crops. The potential ET is also used for forest crops because there is little information on water use of different forest species. Estimated monthly potential ET values are presented for various locations in humid and subhumid climates in Table 5-5.

For perennial forage crops the crop coefficients in Table 5-6 can be used to estimate the ET. For planning purposes the mean ET values will generally suffice. For grasses used for hay, the kc (maximum) values are reached within 6 to 8 d after cutting. The kc value for open water ranges from 1.1 for humid conditions to 1.15 for dry conditions.

Table 5.5: Selected Examples of Monthly Potential Evapotranspiration for Humid and Subhumid Climates (5)

	Inches per Month					
Month	Paris, TX	Central Missouri	Jonesboro, GA	Seabrook, NJ	Hanover, NH	Brevard, NC
Jan	0.6	0.3	0.5	0.1	0.0	0.1
Feb	0.6	0.5	0.5	0.1	0.0	0.1
Mar	1.4	1.2	1.2	0.8	0.0	0.8
Apr	2.7	2.6	2.3	1.6	1.2	1.8
May	4.0	4.3	4.4	3.0	3.3	3.0
June	5.9	5.8	5.9	4.6	5.2	4.1
July	6.4	6.8	6.3	5.6	5.5	4.6
Aug	6.5	6.1	6.0	5.4	4.8	4.2
Sept	3.9	4.1	4.4	4.0	3.0	3.0
Oct	2.6	2.5	2.3	2.0	1.6	1.8
Nov	1.1	1.0	1.0	0.8	0.1	0.6
Dec	0.6	0.4	0.5	0.1	0.0	0.1
Annual	36.3	35.6	35.3	28.1	24.7	24.2

Table 5.6: Crop Coefficients for Perennial Forage Crops (2)

	Condition	
Crop	Humid (light to moderate wind)	Dry (light to moderate wind)
Alfalfa		
Minimum	0.50	0.40
Mean	0.85	0.95
Peak	1.05	1.15
Grass for hay		
Minimum	0.60	0.55
Mean	0.80	0.90
Peak	1.05	1.10
Clover, grass legumes		
Minimum	0.55	0.55
Mean	1.00	1.05
Peak	1.05	1.15
Pasture		
Minimum	0.55	0.50
Mean	0.95	1.00
Peak	1.05	1.10

Kc (minimum) represents conditions just after cutting.
Kc (mean) represents value between cuttings.
Kc (peak) represents conditions before harvesting under dry soil conditions, under wet conditions increase values by 30%.

Prediction of ET

In the absence of ET or pan evaporation data the ET can be predicted from empirical correlations with temperature, humidity, wind, sunshine, and radiation. Over 30 methods have been developed internationally for different agronomic and environmental conditions. Of these, 16 methods were evaluated at 10 different locations (3). Based on accuracy the top 5 methods for estimating ET for different climatic regimes were:

Coastal	Inland-Semiarid to Arid
1. Christiansen	1. Jensen-Haise and van Bavel-Businger, 0.25
2. Turc	2. Penman
3. Kohler	3. Kohler
4. Blaney-Criddle and Ivanov	4. van Bavel-Businger, 0.5
5. Makkink, Penman, and Stephens-Stewart	5. Olivier

On the basis of recommendations made in a publication of the United Nations Food and Agriculture Organization, three methods have potential for widespread use (2). These are the modified Blaney-Criddle method, the Radiation method, and the modified Penman.

The modified Blaney-Criddle method is recommended when only air temperature data are available and is best suited to long periods (1 month or more) of time. In the western U.S. it has been used extensively and is the standard method used by the SCS. In the eastern U.S. it has been less widely used and often produces estimates that are too low (3).

The radiation method is recommended when temperature and radiation or percent cloudiness data are available. Several versions of the method exist, and because they were mainly derived under cool coastal conditions, the resulting ET generally is underestimated (3).

The modified Penman method is one of the most accurate methods when temperature, humidity, wind, and radiation data are available. Along with the radiation method, it offers the best results for periods as short as 10 d.

Other methods exist, such as the Thornthwaite method, in which temperature and latitude are correlated with ET. This method was developed for humid conditions in the east-central U.S. and its application to arid and semi-arid conditions will result in substantial underprediction of ET (3).

AGRONOMIC CROP SELECTION

Varities (cultivars) of major grain, food, and fiber crops are bred specifically for different regions of the United States because of differences in growing seasons, moisture availability, soil type, winter temperatures, and incidence of plant diseases. A regional approach, therefore, is recommended for selection and management of vegetation at land treatment sites (6).

Slow Rate Systems

The crop is an essential component of the SR process for municipal wastewater treatment. In some industrial wastewater SR systems, bare land can be used, particularly if nitrogen removal is unnecessary. The function of the crop in the SR process is to remove nutrients by crop uptake, reduce erosion, and maintain or increase infiltration rates. Crops can also be grown for revenue where local markets are available and the crops are compatible with the wastewater treatment objectives.

Important crop characteristics for SR systems include potential as revenue producer, potential as water user, potential as nitrogen user, and moisture tolerance. Some crops, such as alfalfa, are high water users but cannot tolerate prolonged soil saturation.

Most SR systems are designed to minimize land area by using maximum hydraulic loading rates. Crops that are compatible with high hydraulic loading rates are those having high nitrogen uptake capacity, high consumptive water use, and high tolerance to moist soil conditions. Other desirable crop characteristics for this situation are low sensitivity to wastewater constituents, and minimum management requirements.

Forage and Turf Crops: Forage and turf crops are most compatible with the SR objective of maximum hydraulic loading. Forage crops that have been used successfully include: Reed canarygrass, tall fescue, perennial ryegrass, Italian ryegrass, orchardgrass, and bermudagrass. If forage utilization and value are not a consideration, Reed canarygrass is often a first choice in its area of adaptation because of high nitrogen uptake rate, winter hardiness, and persistence. However, Reed canarygrass is slow to establish and should be planted initially with a companion grass (ryegrass, orchardgrass, or tall fescue) to provide good initial cover.

Of the perennial grasses grown for forage utilization and revenue under high wastewater loading rates, orchardgrass is generally considered to be more acceptable as animal feed than tall fescue or Reed canarygrass. However, orchardgrass is prone to leaf diseases in the southern and eastern states. Tall fescue is generally preferred as a feed over Reed canarygrass but is not suitable for use in the northern tier of states due to lack of winter-hardiness. Other crops may be more suitable for local conditions and advice of local farm advisers or extension specialists will be helpful in making the crop selection.

Turfgrasses are excellent choices for SR systems because they use large amounts of nitrogen and water and use it over much of the year. Golf courses also make good land use candidates for SR system being long term users of irrigation water in most areas. At Tucson, AZ research was conducted on Tifway, giant and common bermuda, overseeded with ryegrass in the winter, and with tall fescue. The Tifway (hybird warm-season bermudagrass) was the choice for irrigation with wastewater (7).

In Florida three varieties of turfgrass-Emerald zoysiagrass, Floratam St. Augustinegrass, and Tifway bermudagrass--were grown using brewery wastewater (8). Similar operations have been established at Houston, TX, and at Fairfield and Bakersfield, CA.

Field Crops: Corn is an attractive crop because of its potentially high rate of economic return as grain or silage. The limited root biomass early in the season and the limited period of rapid nutrient uptake, however, can present problems for nitrogen removal. Prior to the fourth week, root biomass is too low to renovate the wastewater effectively, and after the ninth week, plant uptake slows. During the rapid uptake period, however, corn removes nitrogen efficiently from percolating wastewater (6).

Intercropping is a method of expanding the nutrient and hydraulic capacity of a field corn crop system. A dual system of rye intercropped with corn to maximize the period of nutrient uptake was studied in Michigan and Minnesota (9). For such dual corn-ryegrass cropping systems, rye can be seeded in the standing corn in August, or after the harvest in September. The growth of rye in the spring, before the corn is planted, allows the early application of high nitrogen wastewater. While planting the corn, a herbicide can be applied in strips to kill some rye so that the corn can be seeded in the killed rows. With the remaining rye absorbing nitrogen, less is leached during the early growth of the corn. Alternatively, forage grasses can be intercropped with corn. This "no-till" corn management consists of planting grass in the fall and then applying a herbicide in the spring before planting the corn. When the corn completes its growth cycle, grass is reseeded. Thus, cultivation is reduced; water use is maximized; nutrient uptake is enhanced; and revenue potential is increased.

The most common agricultural crops grown for revenue using wastewater are corn (silage), alfalfa (silage, hay, or pasture), forage grass (silage, hay or pasture), grain sorghum, cotton, and grains. However, any crop, including food crops, may be grown with reclaimed wastewater after suitable preapplication treatment.

In areas with a long growing season, such as California, selection of a double crop is an excellent means of increasing the revenue potential as well as the annual consumptive water use and nitrogen uptake of the crop system. Double crop combinations that are commonly used include (1) short season varieties of soybeans, silage corn, or sorghum as a summer crop; and (2) barley, oats, wheat, vetch, or annual forage grass as a winter crop.

Nutrient Uptake: The highest uptake of nitrogen, phosphorus, and potassium can generally be achieved by perennial grasses and legumes. It should be recognized that whereas legumes normally fix nitrogen from the air, they will preferentially take up nitrogen from the soil-water solution if it is present. The potential for harvesting nutrients with annual crops is generally less than with perennials because annuals use only part of the available growing season for growth and active uptake. Typical annual uptake rates of the major plant nutrients--nitrogen, phosphorus, and potassium--are listed in Table 5-7 for several commonly selected crops.

Table 5.7: Nutrient Uptake Rates for Selected Crops (5)

	lb/acre·yr		
Crop	Nitrogen, N	Phosphorus, P	Potassium, K
Forage crops			
Alfalfa	200-600	20-30	155-200
Bromegrass	115-200	35-50	220
Coastal Bermuda grass	350-600	30-40	200
Kentucky bluegrass	175-240	40	175
Quackgrass	210-250	25-40	245
Reed canarygrass	300-400	35-40	280
Ryegrass	160-250	50-75	240-290
Sweet clover	155	18	90
Tall fescue	130-290	27	270
Orchardgrass	220-310	18-45	200-280
Field crops			
Barley	110	13	18
Corn	155-180	18-27	100
Cotton	65-100	13	36
Grain sorghum	120	13	60
Potatoes	200	18	220-290
Soybeans	220	10-18	27-50
Wheat	140	12	18-50

The nutrient removal capacity of a crop is not a fixed characteristic but depends on the crop yield and the nutrient content of the plant at the time of harvest. Design estimates of harvest removals should be based on yield goals and nutrient compositions that local experience indicates can be achieved with good management on similar soils.

Nitrogen: The rate of nitrogen uptake by crops changes during the growing season and is a function of the rate of dry matter accumulation and the nitrogen content of the plant. Consequently, the pattern of nitrogen uptake is subject to many environmental and management variables and is crop specific. Examples of measured nitrogen uptake rates versus time are shown in Figure 5-3 for annual crops and perennial forage grasses receiving wastewater.

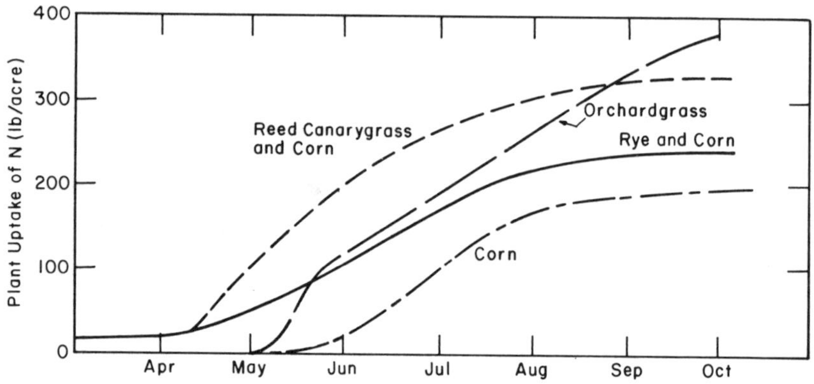

Figure 5-3 Nitrogen uptake versus growing days for annual and perennial crops, After (5).

Some forage crops can have even higher nitrogen uptakes than those in Table 5-7. Californiagrass, a wetland species, widely distributed in the subtropics, was grown with effluent in Hawaii (10). Mean crop yield was 43 tons/acre·yr and nitrogen uptake was 1,870 lb/acre·yr. The nitrogen crop uptake for turfgrasses in Tucson (common bermudagrass overseeded with winter ryegrass) is 525 lb/acre·yr (7).

Example 5-2: Nitrogen Uptake

Conditions: Determine the nitrogen uptake, given an alfalfa yield of 8 tons/acre and a protein content of 20%. The protein content divided by 6.25 gives the nitrogen content.

Solution:

1. $\frac{20\%}{6.25} = 3.2\%$

2. Dry matter of 8 tons/acre = 16,000 lb/acre

3. Nitrogen uptake = 0.032 (16,000)
   = 512 lb/acre

Phosphorus: The amounts of phosphorus in applied wastewater are usually much higher than plant requirements. Fortunately, most soils have a high sorption capacity for phosphorus and very little of the excess passes through the soil.

Potassium: Potassium is used in large amounts by many crops, but typical wastewater is relatively deficient in this element. For example, at 15 mg/L, a typical wastewater contains 40 lb/acre·ft. In many cases, fertilizer potassium (or sludge potassium) may be needed for optimal plant growth depending on the soil and crop. For soils having low levels of natural potassium, a relationship has been developed to estimate potassium loading requirements (5):

$$K_f = 0.9\,U - K_{ww} \tag{5-2}$$

where K_f = annual potassium needed, lb/acre

U = annual crop uptake of nitrogen

K_{ww} = annual wastewater loading of potassium, lb/acre.

Other macronutrients taken up by crops include magnesium, calcium and sulfur; deficiencies of these nutrients are possible in some areas.

The micronutrients important to plant growth (in descending order) are: iron, manganese, zinc, boron, copper, molybdenum, and, occasionally, sodium, silicon, chloride, and cobalt. Most wastewaters contain an ample supply of these elements: in some cases, phytotoxicity may be a consideration.

Overland Flow Systems

A perennial close-growing grass crop is required for overland flow systems. The OF grass crop must have high moisture tolerance, long growing season, and be suited to the local climate.

A mixture of grasses is generally preferred over a single species as shown in Table 5-8. The mixture should contain grasses whose growth characteristics compliment each other, such as sod formers and bunch grasses and species that are dormant at different times of the year.

Table 5.8: Grasses Used at Overland Flow Sites (11, 12)

Site	Type of Grass
Ada, Oklahoma	Annual ryegrass, bermudagrass and Kentucky 31 fescue
Carbondale, Illinois	Tall fescue
Davis, California	Fescue and perennial ryegrass
Easley, South Carolina	Kentucky 31 tall fescue
Hanover, New Hampshire	Orchardgrass, quackgrass, Reed canarygrass, perennial ryegrass
Hunt-Wesson (Davis, California)	Fescue, trefoil, Reed canarygrass
Campbell Soup Co. (Paris, Texas)	Reed canarygrass, redtop, tall fescue
Utica, Mississippi	Reed canarygrass, Kentucky 31 fescue, perennial ryegrass, common bermudagrass

Another advantage of using a mixture of grasses is that, due to natural selection, one or two grasses will often predominate. A successful combination of grasses has been Reed canarygrass, tall fescue, and ryegrass (see Table 5-8). In the south and southwest, dallisgrass, bermudagrass and redtop have also been successful. In northern climates, substitution of orchardgrass for the dallisgrass and redtop is recommended.

At Hanover, NH barnyardgrass invaded the OF slopes and began to dominate the perennial grasses. Being an annual grass, when the barnyardgrass died, it left bare areas that were subject to erosion (13).

Grasses to be avoided include those sensitive to salt (like clover) and those that have long slender seed stalks (Johnson grass and yellow foxtail). In the early stages of development grasses Johnson grass will provide an effective cover, however, with maturity the bottom leaves die off and the habitat for microorganisms becomes reduced.

Rapid Infiltration Systems

Vegetation is generally not used in rapid infiltration systems, but when it is, the use is to maintain high infiltration rates, or to stabilize the soils. At Flushing Meadows, AZ bermudagrass was used in the early research, showing a 25% increase in infiltration rates over bare sand (11).

At Ft. Devens, MA and Whittier Narrows, CA, natural vegetation is used to maintain long-term infiltration. Equipment is kept off these RI sites to avoid soil compaction.

Vegetation for RI systems must be water tolerant and in most cases must be able to withstand several days to a week of inundation.

Bermudagrass, Kentucky bluegrass, Reed canarygrass have been shown to survive inundation for up to 10 days (11)(14).

Silty clay loam and clayey sands are marginal soils for RI systems and use of vegetation with these soils should be investigated. At Brookings, SD the vegetated basins consistently provided the highest infiltration rates over a four year study using silty clay loam soils for RI (15).

The effect of different crops on infiltration is shown in Figure 5-4.

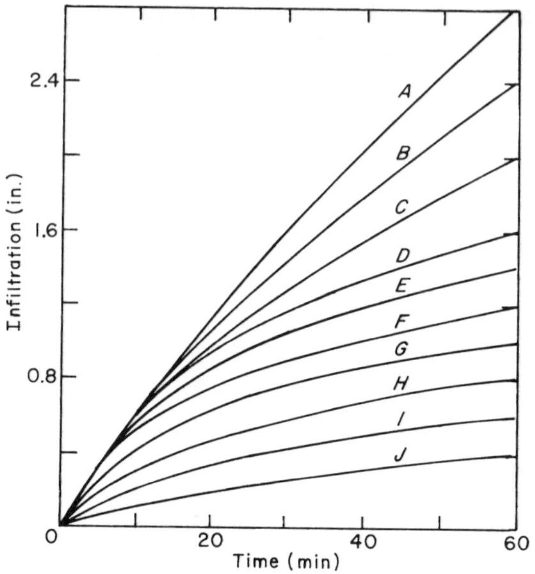

Figure 5-4 Effect of Crop Cover on infiltration rates for various conditions. Symbols A to J represent various conditions: A - old permanent pasture or heavy mulch; B - 4-8 year old permanent pasture; C - 3-4 yr old permanent pasture, light grazing; D - permanent pasture, moderate grazing; E - permanent pasture, heavily grazed; G - Strip crop or mixed cover; H - weeds or grain; I - clean, tilled; J - bare ground, crusted, After (5).

FOREST CROP SELECTION

The most common forest crops used in SR systems have been mixed hardwoods and pines. A summary of representative operational systems and types of forest crops used is presented in Table 5-9. The growth response of trees will vary in accordance with a number of factors; one of the most important is the adaptability of the selected species to the local climate. Local foresters should be consulted for specific recommendations on the likely response of selected species.

Table 5.9: Forested Land Treatment Systems in the U.S.

Location	Design Flow (mgd)	Tree Types
Clayton, Co., GA	19.5	loblolly pines, hardwood
Helen, GA	0.02	mixed pine and hardwood
St. Marys, GA	0.3	slash pine
Mackinaw City, MI	0.2	aspen, birch, white pine
State College, PA	3.0	mixed hardwood, pine
West Dover, VT	0.55	hardwood, balsam, hemlock, spruce

Vegetative uptake and storage of nutrients depend on the species and forest stand density, structure, age, length of season, and temperature. In addition to the trees, there is also nutrient uptake and storage by the understory tree and herbaceous vegetation.

The role of the understory vegetation is particularly important in the early stages of tree establishment.

Forests take up and store nutrients and return a portion of those nutrients back to the soil in the form of leaf fall and other debris such as dead trees. Upon decomposition, the nutrients are released and the trees take them back up. During the initial stages of growth (1 to 2 yr), tree seedlings are establishing a root system; biomass production and nutrient uptake are relatively slow.

To prevent leaching of nitrogen to groundwater during this period, nitrogen loading must be limited or understory vegetation must be established that will take up and store applied nitrogen that is in excess of the tree crop needs.

Nitrogen Uptake

The estimated annual nitrogen uptake of forest ecosystems in selected regions of the United States is presented in Table 5-10. These rates are considered maximum estimates of net nitrogen uptake including both the understory and overstory vegetation during the period of active tree growth.

Table 5.10: Nitrogen Uptake for Selected Forest Ecosystems (5)

	Tree Age (yr)	Average Annual Nitrogen Uptake (lb/acre·yr)
Eastern forests		
Mixed hardwoods	40-60	200
Red pine	25	100
Old field with white spruce plantation	15	200
Pioneer succession	5-15	200
Aspen sprouts	–	100
Southern forests		
Mixed hardwoods	40-60	250
Loblolly pine with no understory	20	200
Loblolly pine with understory	20	250
Lake states forests		
Mixed hardwoods	50	100
Hybrid poplar*	5	140
Western forests		
Hybrid poplar*	4-5	270
Douglas-fir plantation	15-25	200

*Short-term rotation with harvesting at 4 to 5 yr; represents first growth cycle from planted seedlings.

Because nitrogen stored within the biomass of trees is not uniformly distributed among the tree components, the amount of nitrogen that can actually be removed with a forest crop system will be substantially less than the storage estimates given in Table 5-10 unless 100% of the aboveground biomass is harvested (whole-tree harvesting). If only the merchantable stems are removed from the system, the net amount of nitrogen removed by the system will be less than 30% of the amount stored in the biomass.

The distributions of biomass and nitrogen for naturally growing hardwood and conifer (pines, Douglas fir, fir, larch, etc.) stands in temperate regions are shown in Table 5-11. For deciduous species, whole-tree harvesting must take place in the summer when the leaves are on the trees if maximum nitrogen removal is to be achieved.

Table 5.11: Biomass and Nitrogen Distributions by Tree Component for Stands in Temperate Regions (5)

	.... Conifers (%).....		....Hardwoods (%)....	
Tree Component	Biomass	Nitrogen	Biomass	Nitrogen
Roots	10	17	12	18
Stems	80	50	65	32
Branches	8	12	22	42
Leaves	2	20	1	8

Following the initial growth stage, the rates of growth and nutrient uptake increase and remain relatively constant until maturity is approached and the rates decrease. When growth rates and nutrient uptake rates begin to decrease, the stand should be harvested or the nutrient loading decreased. Maturity may be reached at 20 to 25 yr for southern pines, 50 to 60 yr for hardwoods, and 60 to 68 yr for some of the western conifers such as Douglas fir. Of course, harvesting may be practiced well in advance of maturity as with short-term rotation management.

Eastern Forests: During the past 20 yr wastewater has been applied to several forest ecosystems at the Pennsylvania State University (17). Satisfactory renovation was obtained in all systems (eastern mixed hardwoods and red pine) when wastewater was applied during the growing season at 1 in/wk with annual nitrogen loadings of 134 lb/acre. The white spruce/old field forest ecosystem produced a percolate nitrogen concentration of 7.4 mg/L when the hydraulic loading was 2 in/wk and the annual nitrogen loading was 275 lb/acre.

Southern Forests: In a study of a southern mixed hardwood (80% hardwood, 20% pine) forest near Helen, Georgia on a 30% slope with a loading rate of 3 in/wk, about 60% of the applied nitrogen was accounted for in uptake and denitrification. The nitrogen loading was 608 lb/acre and the percolate nitrate concentration was 3.7 mg/L (18).

Lake States Forests: Studies at Michigan State University have shown rather poor nitrogen removal by mature northern hardwoods. Younger forest systems and poplar plantations have shown greater nitrogen uptake, especially during the years when herbaceous cover is present (19).

Western Forests: The wastewater renovation capacity of a newly established plantation of Douglas fir and a mature 50-yr old Douglas fir forest was studied with wastewater nitrogen loadings of 310 to 360 lb/acre·yr (20). The uptake rates, presented in Table 5-10, reflect a substantial uptake by the understory grasses.

Phosphorus and Trace Metals

The assimilative capacity for both phosphorus and trace metals is controlled more by soil properties than plant uptake. The relatively low pH (4.2 to 5.5) of most forest soils is favorable to the retention of phosphorus but not trace metals. However, the high level of organic matter in forest soil improves the metal removal capacity. The amount of phosphorus in trees is small, usually less than 27 lb/acre; therefore, the amount of annual phosphorus accumulation in the biomass is quite small.

CROP MANAGEMENT AND WATER QUALITY

Crop planting, harvesting and pest control are management areas requiring proper techniques to ensure a healthy crop. In addition, wastewaters may have constituents that are harmful to plants (phytotoxic) or that reduce the quality of the crop. Water quality parameters of concern for crop irrigation include salinity, boron, sodium, chloride and pH. Trace elements, particularly zinc, copper, and nickel are of concern for phytotoxicity. However, the concentration of these elements in wastewaters is well below the toxic level of all crops and phytotoxicity could only occur as a result of long-term accumulation of these elements in the soil. See the pertinent sections in Chapter 3 for discussion of all these factors.

Crop Planting and Harvesting: Local extension services or similar experts should be consulted regarding planting techniques and schedules. Most crops require a period of dry weather before harvest to mature and reach a moisture content compatible with harvesting equipment. Soil moisture at harvest time should be low enough to minimize compaction by harvesting equipment. For these reasons, application should be discontinued well in advance of harvest. The time required for drying will depend on the soil drainage and the weather. A drying time of 1 to 2 wk is usually sufficient if there is no precipitation. However, advice on this should be obtained from local agricultural experts.

Harvesting of grass crops and alfalfa involves regular cuttings, and a decision regarding the trade-off between yield and quality must be made. Advice can be obtained from local agricultural experts. In the northeast and north central states, three cuttings per season have been successful with grass crops.

Grazing: Grazing of pasture by beef cattle or sheep can provide an economic return for SR systems. No health hazard has been associated with the sale of the animals for human consumption.

Grazing animals return nutrients to the ground in their waste products. The chemical state (organic and ammonia nitrogen) and rate of release of the nitrogen reduces the threat of nitrate pollution of the groundwater. Much of the ammonia-nitrogen volatilizes and the organic nitrogen is held in the soil where it is slowly mineralized to ammonium and nitrate forms (See Chapter 3).

In terms of pasture management, cattle or sheep must not be allowed on wet fields to avoid severe soil compaction and reduced soil infiltration rates. Wet grazing conditions can also lead to animal hoof diseases. Pasture rotation should be practiced so that wastewater can be applied immediately after the livestock are removed. In general, a pasture area should not be grazed longer than 7 days. Typical regrowth periods between grazings range from 14 to 36 days. Depending on the period of regrowth provided, one to three water applications can be made during the regrowth period. Rotation grazing cycles for 2 to 8 pasture areas are given in Table 5-12. At least 3 to 4 days drying time following an application should be allowed before livestock are returned to the pasture.

Table 5.12: Pasture Rotation Cycles for Different Numbers of Pasture Areas

Number of Pasture Areas	Rotation Cycle (days)	Regrowth Period (days)	Grazing Period (days)
2	28	14	14
3	30	20	10
4	28	21	7
5	35	28	7
6	36	30	6
7	42	36	6
8	40	35	5

Agricultural Pest Control: Problems with weeds, insects, and plant diseases are aggravated under conditions of frequent water application, particularly when a single crop is grown year after year or when no-till practices are used. Most pests can be controlled by selecting resistant or tolerant crop varieties and by using pesticides in combination with appropriate cultural practices. State and local experts should be consulted in developing an overall pest control program for a given situation.

Overland Flow Crop Management

After the cover crop has been established, the OF slopes will need little, if any, maintenance work. It will, however, be necessary to mow the grass periodically. A few systems have been operated without cutting, but the tall grass tends to interfere with maintenance operations. Normal practice has been to cut the grass two or three times a year. The first cutting may be left on the slopes. After that, however, it is desirable to remove the cut grass. The advantages of doing so are that additional nutrient removal is achieved, channeling problems may be more readily observed, and revenue can sometimes be produced by the sale of hay. Depending on the local market conditions, the cost of harvesting can at least be offset by the sale of hay (5).

Slopes must be allowed to dry sufficiently such that mowing equipment can be operated without leaving ruts or tracks that will later result in channeling of the flow. The drying time required before mowing varies with the soil and climatic conditions and can range from a few days to a few weeks. The downtime required for harvesting can be reduced by a week or more if green-chop harvesting is practiced instead of mowing, raking, and baling. However, local markets for green-chop must exist for this method to be feasible.

It is common for certain native grasses and weeds to begin growing on the slopes. Their presence usually has little impact on treatment efficiency and it is generally not necessary to eliminate them. However, there are expectations and the local extension services should be consulted for advice.

Proper management of the slopes and the application schedule will prevent conditions conducive to mosquito breeding. Other insects are usually no cause for concern, although an invasion of certain pests such as army worms may be harmful to the vegetation and may require periodic insecticide application.

<u>Forest Crop Management</u>

The type of forest crop management practice selected is determined by the species mix grown, the age and structure of the stand, the method of reproduction best suited and/or desired for the favored species, terrain, and type of equipment and technique used by local harvesters. The most typical forest management situations encountered in land treatment are management of existing forest stands, reforestation, and short-term rotation.

<u>Existing forests</u>: The general objective of the forest management

program is to maximize biomass production. The compromise between fully attaining a forest's growth potential and the need to operate equipment efficiently (distribution and harvesting equipment) requires fewer trees per unit area. These operations will assure maintenance of a high nutrient uptake by the forest.

In even-aged forests, trees will all reach harvest age at the same time. The usual practice is to clear-cut these forests at harvest age and regenerate a stand by either planting seedlings, sprouting from stumps (called coppice), or a combination of several of the methods. Even-aged stands may require a thinning at an intermediate age to maintain maximum biomass production. Coniferous forests, in general, must be replanted, whereas hardwood forests can be reproduced by coppice or natural seeding.

For uneven-aged forests, the desired forest composition, structure, and vigor can be best achieved through thinning and selective harvest. However, excessive thinning can make trees susceptible to wind throw and caution is advised in windy areas. The objectives of these operations would be to maintain an age class distribution in accordance with the concept of optimum nutrient storage. The maintenance of fewer trees than normal would permit adequate sunlight to reach the understory to promote reproduction and growth of the understory. Thinning should be done initially prior to construction of the distribution system and only once every 10 years or so to minimize soil and site damage.

The concept of "whole-tree harvesting" should be considered for all harvesting operations, whether it be thinning, selection harvest, or clear-cut harvest. Whole-tree harvesting removes the entire standing tree: stem, branches, and leaves. Thus, 100% of the nitrogen accumulated in the aboveground biomass would be removed.

Prescribed fire is a common management practice in many forests to reduce the debris or slash left on the site during conventional harvesting methods. During the operation, a portion of the forest floor is burned and nitrogen is volatilized. Although this represents an immediate benefit in terms of nitrogen removal from the site, the buffering capacity that the forest floor offers is reduced and the likelihood of a nitrate leaching to the groundwater is increased when application of wastewater is resumed.

Reforestation: Wastewater nutrients often stimulate the growth of the herbaceous vegetation to such an extent that they compete with and shade out the desirable forest species. Herbaceous vegetation is necessary to act as a nitrogen sink while the trees are becoming established, and therefore, cultural practices must be designed to control but not eliminate the herbaceous vegetation. As the tree crowns begin to close, the herbaceous vegetation will be shaded and its role in the renovation cycle reduced. Another alternative to control of the herbaceous vegetation is to eliminate it completely and reduce the hydraulic and nutrient loading during the establishment period.

Short-Term Rotation: Short-term rotation forests are plantations of closely spaced hardwood trees that are harvested repeatedly on cycles of less than 10 yr. The key to rapid growth rates and biomass development is the rootstock that remains in the soil after harvest and then resprouts. Short-term rotation harvesting systems are readily mechanized because the crop is uniform and relatively small.

Using conventional tree spacings of 8 to 12 ft, research on systems where wastewater has been applied to short-term rotation plantations has shown that high growth rates and high nitrogen removal are possible (5). Planted stock will produce only 50% to 70% of the biomass produced following cutting and resprouting (5). If nitrogen and other nutrient uptake is proportional to biomass, the first rotation from planted stock will not remove as much as subsequent rotations from coppice. Therefore, the initial rotation must receive a reduced nutrient load or other herbaceous vegetation must be employed for nutrient storage. Alternatively, closer tree spacings may be used to achieve desired nutrient uptake rates during initial rotation.

REFERENCES

1. Rosenberg, N.J. Microclimate: The Biological Environment, New York: John Wiley & Sons (1974).

2. Doorenbos, J. and Pruitt, W.O. Crop Water Requirements. FAO Irrigation and Drainage Paper No. 24. Food and Agricultural Organization of the United Nations, Rome (1977).

3. Jensen, M.E., et al. Consumptive Use of Water and Irrigation Water Requirements. ASCE Committee on Irrigation Water Requirements (September 1973).

4. California Department of Water Resources. Vegetative Water Use in California, 1974. Bulletin No. 113-3. Sacramento, California (April 1975).

5. Process Design Manual for Land Treatment of Municipal Waste-Water, U.S. Environmental Protection Agency, EPA 625/1-81-013 (October 1981).

6. D'Itri, F.M. Land Treatment of Municipal Wastewater: Vegetation Selection and Management, Ann Arbor, Michigan: Ann Arbor Science (1982).

7. Pepper, I.L. Land Application of Municipal Effluent on Turf. In: Proceedings of the 1981 Technical Conference Silver Spring, Maryland: The Irrigation Association (1981).

8. White, R.W. Utilizing Food Processing Wastewater for Turf Production. In: Proceedings of the 1980 Technical Conference Silver Spring, Maryland: The Irrigation Association (1980).

9. Brockway, D.G., et al. The Current Status on the Selection and Management of Vegetation for Slow Rate and Overland Flow Application Systems to Treat Municipal Wastewater in the North Central Region of the United States. In: Land Treatment of Municipal Wastewater, D'Itri, F.M. (Ed.) Ann Arbor, Michigan: Ann Arbor Science, pp. 5-18 (1982).

10. Handley, L.L. Effluent Irrigation of Californiagrass. Proceedings Water Reuse Symposium II, Vol. 2, AWWA Research Foundation, Washington, D.C. (August 23-28, 1981).

11. Pound, C.E. and Crites, R.W. Wastewater Treatment and Reuse by Land Application, Vol. II, Environmental Protection Agency, EPA-660/2-73-006b (August 1973).

12. Hinrichs, D.J., et al. Assessment of Current Information on Overland Flow Treatment of Municipal Wastewater. U.S. Environmental Protection Agency, EPA 430/9-80-002, MCD-66 (May 1980).

13. Palazzo, A.J., Jenkins, T.F. and Martel, C.J. Vegetation Selection and Management for Overland Flow Systems. In: Land Treatment of Municipal Wastewater, D'Itri, F.M. (Ed.) Ann Arbor, Michigan: Ann Arbor Science, pp. 135-154 (1982).

14. Erbisch, F.H. and Stark, K.L. Effects of Inundation on Six Varieties of Turfgrass, U.S. Army Corps of Engineers, CRREL Special Report 82-12 (May 1982).

15. Dornbush, J.N. Infiltration Land Treatment of Stabilization Pond Effluent. U.S. Environmental Protection Agency. EPA-600/2-81-226 (September 1981).

16. US EPA, Process Design Manual-Land Application of Municipal Sludge, EPA-625/1-83-016 (Oct. 1983).

17. Sopper, W.E. and Kerr, S.N. Renovation of Municipal Wastewater in Eastern Forest Ecosystems. In: Utilization of Municipal Sewage Effluent and Sludge on Forest and Disturbed Land, Sopper, W.E. and Kerr, S.N. (Eds.) University Park, Pennsylvania: The Pennsylvania State University Press, pp. 61-76 (1979).

18. Nutter, W.L., Schultz, R.C. and Brister, G.H. Land Treatment of Municipal Wastewater on Steep Forest Slopes in the Humid Southeastern United States. Proceedings of the International Symposium on Land Treatment of Wastewater, Vol. 1, Hanover, New Hampshire, pp. 265-274 (1978).

19. McKim, H.L., et al. Wastewater Applications in Forest Ecosystems. U.S. Army Corps of Engineers. CRREL Report 82-19. (August 1982).

20. Cole, D.W. and Schiess, P. Renovation of Wastewater and Response of Forest Ecosystems: the Pack Forest Study. Proceedings of the International Symposium on Land Treatment of Wastewater. Vol. 1, Hanover, New Hampshire, pp. 323-331 (1978).

6

Site Identification and Selection

Process and site selection in land treatment are interrelated. The ability of the land treatment processes to remove wastewater constituents, described in Chapter 3, the discharge quality criteria, and the soil and other site characteristics affect the choice of the appropriate land treatment process. The presence of a suitable site within an economical transmission distance from the wastewater source will determine if a land treatment system can be implemented.

Because the selection of a process and site for land treatment are related, a 2-phased planning procedure is often used. The two phases are presented in Figure 6-1.

The first phase involves estimating preliminary land area requirements based on wastewater and climate characteristics, identifying potential sites in the area, evaluating the sites based on technical and economic factors, and selecting potential sites.

The second phase, assuming sites are selected, involves field investigations, preliminary design and cost estimates, comparison to other alternatives, and selection of the most economical alternative.

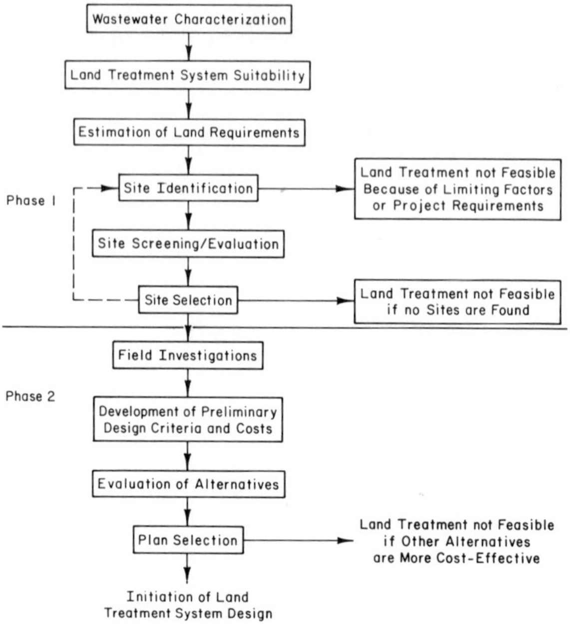

Figure 6-1 Two phase planning process.

PRELIMINARY LAND REQUIREMENTS

Preliminary land requirements can be estimated for each land treatment process, based on wastewater characteristics and climatic conditions. Wastewater characteristics include average annual flows and concentrations of constituents such as BOD, suspended solids, nitrogen, phosphorus and trace elements.

Wastewater Characteristics

Municipal wastewater flows range typically from 65 to 100 gallons per capita per day (gpcd). Industrial wastewater flows are too variable to generalize and must be estimated from information specific to the product and wastewater producing operations. Existing wastewater flow records or water use records should be used where available.

Constituent concentrations that are seen typically in municipal wastewater are presented in Table 6-1. These characteristics represent medium strength wastewater.

Table 6.1: Typical Characteristics of Municipal Wastewater

Constituent	Concentration (mg/ℓ)
BOD	200
Suspended solids	200
Nitrogen, total	30
Organic nitrogen	15
Ammonia nitrogen	15
Phosphorus, total	10
Potassium	15

Industrial wastewaters vary widely in their characteristics, especially for organics, metals, and nitrogen. Characteristics of food processing wastewaters that have been applied directly to the land are presented in Table 6-2. Wastewater characterization is necessary in planning for industrial land application systems.

Table 6.2: Characteristics of Food Processing Wastewaters Applied to the Land (1)

Constituent	Concentration (mg/ℓ)*
BOD	200-33,000
Suspended solids	200-3,000
Total fixed dissolved solids	<1,800
Total nitrogen	10-1,900
pH, units	3.5-12.0
Temperature, °C	<65

*Except as noted.

Preliminary Loading Rates

In the absence of site information, typical loading rates can be assumed to initiate the planning process. For SR systems the degree of preapplication treatment (either primary or secondary) has little affect on the loading rate. For OF and RI systems, higher loading rates can usually be used with higher quality effluent. Typical loading rates for preliminary estimates of land requirements are presented in Table 6-3.

The rates in Table 6-3 are necessarily conservative. Once a potential site has been analyzed and the ability to meet discharge requirements is assessed, the loading rates can usually be increased.

Table 6.3: Preliminary Loading Rates for Initial Estimate of Land Requirements

Process	Loading Rate (inches/week)
Slow rate	
Agricultural.	1.5
Forest	1.0
Rapid infiltration	
Primary effluent	12
Secondary effluent	20
Overland flow	
Screened wastewater and primary effluent	4
Secondary effluent	8

Storage Needs

Storage of wastewater may be necessary due to cold weather, excessive precipitation, or crop management. For preliminary estimates it is usually sufficient to base storage needs on climatic factors. A map showing storage days based on cold weather and excessive precipitation is presented in Figure 6-2. This figure should be used for a preliminary estimate of storage needed for OF systems. For SR systems using agricultural crops, the crop management time for harvesting and planting should be added to the storage days taken from Figure 6-2. The values in Figure 6-2 are not valid for RI and forested SR systems, since both can be operated during subfreezing weather. For RI and forested SR systems, a minimum storage of 7 to 14 days can be assumed for preliminary estimates of land area.

Site Area Estimate

Preliminary site area requirements can be estimated from wastewater flows, storage needs, and preliminary loading rates. The relationship between field area, loading rates, and operating period is shown in Equation 6-1.

$$F = \frac{13,443\ Q}{L\ P} \qquad (6\text{-}1)$$

where F = field area, acres

Q = average flow, mgd

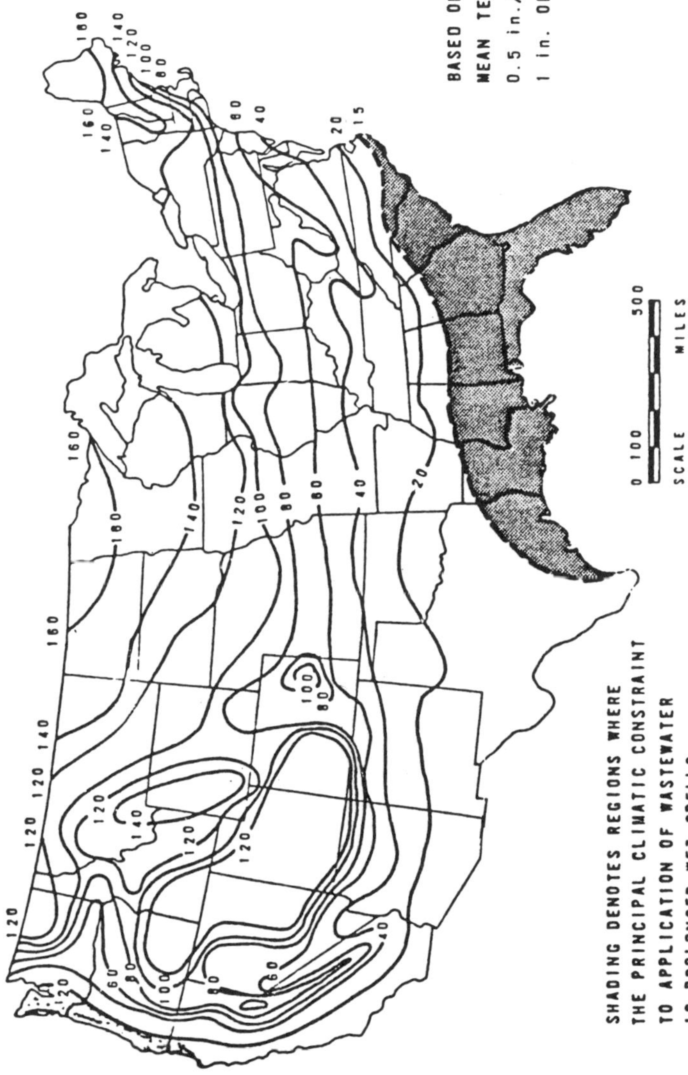

Figure 6-2 Estimated storage days based on climatic factors only.

L = loading rate, in./wk

P = period of application, wk/yr

$$13{,}443 = \text{conversion factor} = 3.069\frac{\text{acre-ft}}{\text{mgd}} \times \frac{12\ \text{in.}}{\text{ft}} \times \frac{365\ \text{d}}{\text{yr}}$$

The period of application can be approximated by subtracting the estimated storage period from 52 wk/yr. Site areas for a 1 mgd flow for all three systems are presented in Table 6-4. For SR and RI systems the numbers in Table 6-4 include 20% extra area over the calculated field area to account for unusable land. For OF systems the extra land in Table 6-4 is 40% to account for the additional inefficiency in constructing OF slopes.

Table 6.4: Site Identification Land Requirements

System	Land Requirements (acres/mgd)
Slow rate, agricultural	
No storage	200
1 month's storage	225
2 months' storage	250
3 months' storage	275
4 months' storage	315
5 months' storage	350
6 months' storage	415
Slow rate, forest	
No storage	310
1 month's storage	335
Rapid infiltration	
Primary effluent	30
Secondary effluent	15

Overland flow

Storage (months)	Screened Wastewater	Secondary Effluent
0	90	180
1	100	200
2	110	220
3	120	240
4	140	280

SITE IDENTIFICATION

To identify potential land treatment sites it is necessary to obtain data on land use, soil types, and topography. The types and sources of data needed to identify and evaluate potential sites are presented in Table 6-5.

Table 6.5: Types and Sources of Data Required for Land Treatment Site Evaluation

Type of Data	Principal Source
Topography	USGS topographic quads
Soil type and permeability	SCS soil survey
Temperature (mean monthly and growing season)	SCS soil survey, NOAA, local airports, newspapers
Precipitation (mean monthly, maximum monthly)	SCS soil survey, NOAA, local airports, newspapers
Evapotranspiration and evaporation (mean monthly)	SCS soil survey, NOAA, local airports, newspapers, agricultural extension service
Land use	SCS soil survey, aerial photos from the Agricultural Stabilization and Conservation Service, and county assessors' plats
Zoning	Community planning agency, city or county zoning maps
Agricultural practices	SCS soil survey, agricultural extension service, county agents
Surface and groundwater discharge requirements	State or EPA
Groundwater (depth and quality)	State water agency, USGS, driller's logs of nearby wells

Use of Map Overlays

The complexity of site identification depends on the size of the study area and the nature of the land use. One approach is to start with land use plans and identify undeveloped land. Map overlays can then help the planner or engineer to organize and study the combined effects of land use, slope, relief, and soil permeability. Criteria can be set on these four factors, and areas that satisfy the criteria can then be located. If this procedure is used as a preliminary step in site identification, the criteria should be reassessed during each successive iteration. Otherwise, strict adherence to such criteria may result in overlooking either sites or land treatment opportunities.

Site Suitability Factors

Potential land treatment sites are identified using a deductive approach. First, any constraints that might limit site suitability are identified. In most study areas, all land within the area should be evaluated for each land treatment process. The next step is to classify broad areas of land near the area where wastewater is generated according to their land treatment suitability. Factors that should be considered include current and planned land use, topography, soils, geology, groundwater and surface water hydrology.

Land Use: Land use in most communities is regulated by local, county and regional zoning laws. Land treatment systems must comply with the appropriate zoning regulations. For this reason, the planner should be fully aware of the actual land uses and proposed land uses in the study area. The planner should attempt to develop land treatment alternatives that conform to local land use goals and objectives.

Land treatment systems can conform with the following land use objectives:

- Protection of open space that is used for land treatment
- Production of agricultural or forest products using renovated water on the land treatment site
- Reclamation of land by using renovated water to establish vegetation on scarred land
- Augmentation of parklands by irrigating such lands with renovated water
- Management of flood plains by using flood plain areas for land treatment, thus precluding land development on such sites
- Formation of buffer areas around major public facilities, such as airports

To evaluate present and planned land uses, city, county and regional land use plans should be consulted. Because such plans often do not reflect actual current land use, site visits are recommended to determine existing land use. Aerial photographic maps may be obtained from the Soil Conservation Service (SCS) or the local assessor's office. Other useful information may be available from the USGS and the EPA, including true color, false color infrared, and color infrared aerial photos of the study area.

Once the current and planned land uses have been determined, they should be plotted on a study area map. Then, land use suitability may be plotted using the factors shown in Table 6-6.

Table 6.6: Land Use Suitability Factors for Identifying Land Treatment Sites (3)

	Type of System			
Land Use Factor	Agricultural Slow Rate	Forest Slow Rate	Overland Flow	Rapid Infiltration
Open or cropland	High	Moderate	High	High
Partially forested	Moderate	Moderately high	Moderate	Moderate
Heavily forested	Low	High	Low	Low
Built upon (residential, commercial, or industrial)	Low	Very low	Very low	Very low

Both land acquisition procedures and treatment system operation are simplified when few land parcels are involved and contiguous parcels are used. Therefore, parcel size is an important parameter. Usually, information on parcel size can be obtained from county assessor or county recorder maps. Again, the information should be plotted on a map of the study area.

Topography: Steep grades limit a site's potential because the amount of runoff and erosion that will occur is increased, crop cultivation is made more difficult, if not impossible, and saturation of steep slopes may lead to unstable soil conditions. The maximum acceptable grade depends on soil characteristics and the land treatment process used.

Grade and elevation information can be obtained from USGS topographic maps, which usually have scales of 1:24,000 (7.5 minute series) or 1:62,500 (15 minute series). Grade suitability may be plotted using the criteria listed in Table 6-7.

Table 6.7: Grade Suitability Factors for Identifying Land Treatment Sites (3)

Grade Factor	Slow Rate Systems		Overland	Rapid
(%)	Agricultural	Forest	Flow	Infiltration
0-12	High	High	High	High
12-20	Low	High	Moderate	Low
20	Very low	Moderate	Eliminate	Eliminate

Relief is another important topographical consideration and is the difference in elevation between one part of a land treatment system and another. The primary impact of relief is its effect on the cost of conveying wastewater to the land application site. Often, the economics of pumping wastewater to a nearby site must be compared with the cost of constructing gravity conveyance to more distant sites.

A site's susceptibility to flooding also can affect its desirability. The flooding hazard of each potential site should be evaluated in terms of both the possible severity and frequency of flooding as well as the areal extent of flooding. In some areas, it may be preferable to allow flooding of the application site provided offsite storage is available. Further, crops can be grown in flood plains if flooding is infrequent enough to make farming economical.

Overland flow sites can be located in flood plains provided they are protected from direct flooding which could erode the slopes. Backwater from flooding, if it does not last more than a few days, should not be a problem. Flood plain sites for RI basins should be protected from flooding by the use of levees.

Summaries of notable floods and descriptions of severe floods are published each year as the USGS Water Supply Papers. Maps of certain areas inundated in past floods are published as Hydrologic Investigation Atlases by the USGS. The USGS also has produced more recent maps of flood prone areas for many regions of the country as part of the Uniform National Program for Managing Flood Losses. These maps are based on standard 7.5 minute (1:24,000) topographic sheets and identify areas that lie within the 100 year flood plain. Additional information on flooding susceptibility is available from local offices of the U.S. Army Corps of Engineers and local flood control districts.

Soils: Common soil-texture terms and their relationship to the SCS textural class names are listed in Table 6-8.

Table 6.8: Soil Textural Classes and General Terminology Used in Soil Descriptions

.........General Terms.........		Basic Soil Textural
Common Name	Texture	Class Names
Sandy soils	Coarse	Sand, loamy sand
	Moderately coarse	Sandy loam, fine sandy loam
Loamy soils	Medium	Very fine sandy loam, loam, silt loam, silt
	Moderately fine	Clay loam, sandy clay loam, silty clay loam
Clayey soils	Fine	Sandy clay, silty clay, clay

Fine-textured soils do not drain well and retain water for long periods of time. Thus, infiltration is slower and crop management is more difficult than for freely drained soils such as loamy soils. Fine-textured soils are best suited for the OF process. Loamy or medium-textured soils are desirable for the SR process, although sandy soils may be used with certain crops that grow well in rapidly draining soils. Soil structure and soil texture are important characteristics that relate to permeability and acceptability for land treatment. Structure refers to the degree of soil particle aggregation. A well structured soil is generally more permeable than unstructured material of the same type. The RI process is suited for sandy or loamy soils.

Soils surveys are usually available from the SCS. Soil surveys normally contain maps showing soil series boundaries and textures to a depth of about 5 ft. In a survey, limited information on chemical properties, grades, drainage, erosion potential, general suitability for locally grown crops, and interpretive and management information is provided. Where published surveys are not available, information on soil characteristics can be obtained from the SCS, through the local county agent.

Although soil depth, permeability, and chemical characteristics significantly affect site suitability, data on these parameters are often not available before the site investigation phase. If these data are available, they should be plotted on a study area map along with soil texture. In identifying potential sites, the planner should keep in mind that adequate soil depth is needed for root development and for thorough wastewater treatment. Further, permeability requirements vary among the land treatment processes. Desirable permeability ranges are shown by process in Table 6-9 together with desired soil texture. The SCS permeability class definitions are also shown on Figure 4-6.

Table 6.9: Typical Soil Permeabilities and Textural Classes for Land Treatment Processes

	Land Treatment Processes		
	Slow Rate	Rapid Infiltration	Overland Flow
Soil permeability range, inchs per hour	0.06-20	>2.0	<0.2
Permeability class range	Moderately slow to moderately rapid	Rapid	Slow
Textural class range	Clay loams to sandy loams	Sandy and sandy loams	Clays and clay loams
Unified soil classification	GM-d, SM-d, ML, OL, MH, PT	GW, GP, SW, SP	GM-u, GC, SM-u SC, CL, OL, CH, OH

<u>Geology</u>: Certain geological formations are of interest during Phase 1 investigations. Discontinuities and fractures in bedrock may cause shortcircuiting or other unexpected groundwater flow patterns. Impermeable or semipermeable layers of rock, clay, or hardpan can result in perched groundwater tables. The USGS and many state geological surveys have maps indicating the presence and effects of geological formations. These maps and other USGS studies may be used to plot locations within the study area where geological formations may limit the suitability for land treatment.

<u>Groundwater:</u> A knowledge of the regional groundwater conditions is particularly important for potential rapid infiltration and slow rate sites. Overland flow will not usually require an extensive hydrogeologic investigation. Sufficient removal of pollutants in the applied wastewater before reaching a permanent groundwater resource is the primary concern. The depth to groundwater and its seasonal fluctuation are a measure of the aeration zone and the degree of renovation that will take place.

When several layers of stratified groundwater underlie a particular site, the occurrence of the vertical leakage between layers should be evaluated. Direction and rate of groundwater flow and aquifer permeability together with groundwater depth are useful in predicting the effect of applied wastewater on the groundwater regime. The extent of recharge mounding, interconnection of aquifers, perched water tables, the potential for surfacing groundwater, and the design of monitoring and withdrawal wells are dependent on groundwater flow data.

Much of the data required for groundwater evaluation may be determined through use of existing wells. Wells that could be used for monitoring should be listed and their relative location described. Historical data on quality, water levels, and quantities pumped from the operation of existing wells may be of value. Such data include seasonal groundwater-level variations, as well as variations over a period of years. The USGS maintains a network of about 15,800 observation wells to monitor water levels nationwide. Records of about 3,500 of these wells are published in Water-Supply Paper Series, "Groundwater-Levels in the United States." Many local, regional, and state agencies compile drillers' boring logs that are also valuable for defining groundwater hydrology.

Land treatment of wastewater can provide an alternative to discharge of conventionally treated wastewater. However, the adverse impact of percolated wastewater on the quality of the groundwater must also be considered. Existing groundwater quality should be determined and compared to quality standards for its current or intended use. The expected quality of the renovated wastewater can then be compared to determine which constituents in the renovated water might be limiting. The USGS "Groundwater Data Network" monitors water quality in observation wells across the country. In addition, the USGS undertakes project investigations or areal groundwater studies in cooperation with local, state, or other federal agencies to appraise groundwater quality. Such reports may provide a large part of the needed groundwater data.

Surface Water Hydrology: Surface water hydrology is of interest in land treatment processes mostly because of the runoff of stormwater. Considerations relating to surface runoff control apply to both slow rate and overland flow. Rapid infiltration processes are designed for no runoff.

The control of stormwater runoff both onto and off a land treatment site must be considered. First, the facilities constructed as part of the treatment system must be protected against erosion and washout from extreme storm events. For example, where earthen ditches and/or terraces are used, erosion control from stormwater runoff must be provided. The degree of control of runoff to prevent the destruction of the physical system should be based on the economics of replacing equipment and structures. There is no standard extreme storm event in the design of drainage and runoff collection systems, although a 10 year return event is suggested as a minimum. See Chapter 11 for further discussion of storm water runoff of overland flow sites.

CLIMATIC FACTORS

Local climate may affect (1) the water balance (and thus the acceptable wastewater hydraulic loading rate), (2) the length of the growing season, (3) the number of days per year that a land treatment system cannot be operated, (4) the storage capacity requirement, (5) the loading cycle of RI systems, and (6) the amount of stormwater runoff. For this reason, local precipitation, evapotranspiration, temperature, and wind values must be determined before design criteria can be established. Whenever possible, at least 10 years of data should be used to obtain these values.

Three publications of the National Oceanic and Atmospheric

Administration (NOAA) provide sufficient data for most communities. The Monthly Summary of Climatic Data provides basic information, including total precipitation, temperature maxima and minima, and relative humidity, for each day of the month and every weather station in a given area. Whenever available, evaporation data are included. An annual summary of climatic data, entitled Local Climatological Data, is published for a small number of major weather stations. Included in this publication are the normals, means, and extremes of all the data on record to date for each station. The Climate Summary of the United States provides 10 year summaries of the monthly climatic data. Other data included are:

- Total precipitation for each month of the 10 year period.
- Mean number of days that precipitation exceeded 0.10 and 0.50 in during each month.
- Total snowfall for each month of the period.
- Mean temperature for each month of the period.
- Mean daily temperature maxima and minima for each month.
- Mean number of days per month that the temperature was less or equal to 32°F or greater than or equal to 90°F.

A fourth reference that can be helpful is EPA's Annual and Seasonal Precipitation Probabilities (5). This publication includes precipitation probabilities for 93 stations throughout the United States. Data requirements for planning purposes are summarized in Table 6-10.

Table 6.10: Summary of Climatic Analyses

Factor	Date Required	Analysis	Use
Precipitation	Annual average, maximum, minimum	Frequency	Water balance
Rainfall storm	Intensity, duration	Frequency	Runoff estimate
Temperature	Days with average below freezing	Frost free period	Storage, treatment efficiency, crop growing season
Wind	Velocity, direction	–	Cessation of sprinkling
Evapotranspiration	Annual, monthly average	Annual distribution	Water balance

WATER RIGHTS AND POTENTIAL CONFLICTS

Land application of wastewaters may cause several changes in drainage and flow patterns (6):

1. Site drainage may be affected by land preparation, soil characteristics, slope, method of wastewater application, cover crops, climate, buffer zones, and spacing of irrigation equipment.
2. Land application may alter the pattern of flow in the body of water that would have received the wastewater discharge. Although this may diminish the flow in the body of water, it also may increase the quality. The change may be continuous or seasonal.
3. Land application may cause surface water diversion, because wastewaters that previously would have been carried away by surface waters are now applied to land and often diverted to a different watershed.

Two basic types of water rights laws exist in the United States: Riparian laws, which emphasize the right of riparian landowners along a watercourse to use of the water, and appropriative laws, which emphasize the right of prior users of the water (6). Most riparian or land ownership rights are in effect east of the Mississippi, whereas most appropriative rights are in effect west of the Mississippi River.

Most states divide their water laws into three categories: (1) waters in well-defined channels or basins (natural watercourses), (2) superficial waters not in channels or basins (surface waters), and (3) underground waters not in well-defined channels or basins (percolating waters or groundwaters).

The state or local water master or water rights engineer should be consulted to avoid potential problems. Other references to consider are the publications, A Summary-Digest of State Water

Laws, available from the National Water Commission (5), and Land Application of Wastewater and State Water Law, Volumes I and II (7,8). If problems develop or are likely with any of the feasible alternatives, a water rights attorney should also be consulted.

SITE SELECTION

Once the data on site characteristics are collected and mapped, the site evaluation and selection process can proceed. If the number of sites are few and their relative suitability clearly apparent, a simple economic comparison will lead to selection of the best site. If a number of sites are to be compared, a site screening procedure can be used.

Site Screening Procedure

The general procedure for site suitability rating can be used to compare different sites or it can be used to screen a large site that may have portions suitable to different land treatment processes. A procedure incorporating economic factors is presented for RI and OF systems. A procedure specific to SR forest systems is also included.

The general procedure is to assign numerical values to various site characteristics, with larger numbers indicating highest suitability. The individual numbers for each site or subarea are then added together to obtain the overall suitability rating. The rating factors in Table 6-11 are applicable to all processes. Site selection factors and weightings should vary to suit the needs of the local area and type of sites available.

Example 6-1: Site Suitability Rating

Conditions: Compare the suitabilities of three sites being considered for RI. The characteristics of the three sites are given in Table 6-12.

Table 6.11: Rating Factors for Site Selection (9)

Characteristic	Slow Rate Systems Agricultural	Slow Rate Systems Forest	Overland Flow	Rapid Infiltration
Soil depth, ft*				
1-2	E**	E	0	E
2-5	3	3	4	E
5-10	8	8	7	4
>10	9	9	7	8
Minimum depth to groundwater, ft				
<4	0	0	2	E
4-10	4	4	4	2
>10	6	6	6	6
Permeability, inches/hour***				
<0.06	1	1	10	E
0.06-0.2	3	3	8	E
0.2-0.6	5	5	6	1
0.6-2.0	8	8	1	6
>2.0	8	8	E	9
Grade, %				
0-5	8	8	8	8
5-10	6	8	5	4
10-15	4	6	2	1
15-20	0	5	E	E
20-30	0	4	E	E
30-35	E	2	E	E
>35	E	0	E	E
Existing or planned land use				
Industrial	0	0	0	0
High density residential/urban	0	0	0	0
Low density residential/urban	1	1	1	1
Forested	1	4	1	1
Agricultural or open space	4	3	4	4
Overall suitability rating†				
Low	<15	<15	<16	<16
Moderate	15-25	15-25	16-25	16-25
High	25-35	25-35	25-35	25-35

Note: The higher the maximum number in each characteristic, the more important the characteristic; the higher the ranking, the greater the suitability.

*Depth of the profile to bedrock.
**Excluded; rated as poor.
***Permeability of most restrictive layer in soil profile.
†Sum of values.

Table 6.12: Site Characteristics for Example 6-1

Characteristics	Site 1	Site 2	Site 3
Soil depth, ft	>10	5-10	2-5
Depth to groundwater, ft	>10	>10	4-10
Permeability, inches/hour	>2	0.6-2.0	>2
Grade, %	0-5	0-5	0-5
Land use	Forested	Agricultural	Industrial

Solution: Assign the numerical ratings to each site characteristic using the values in Table 6-11. The assigned numbers are shown in Table 6-13. Based on these five characteristics, Site 1 is the preferred site. Site 2 should be retained for consideration, although its permeability rating makes it less suitable than Site 1. Site 3 should be eliminated because of inadequate soil depth.

Table 6.13: Site Comparison for Example 6-1

	Rating Values		
Characteristics	Site 1	Site 2	Site 3
Soil depth	8	4	E
Depth to groundwater	6	6	2
Permeability	9	6	9
Grade	8	8	8
Land use	1	4	0
Total	32	28	19 (E)
Rating	High	High	Eliminate

Screening Procedure with Economic Factors: In addition to the technical factors listed in Table 6-11 the economics of site development are often critical. These include distance from the wastewater source, elevation differences and the costs for land acquisition and management. Table 6-14 presents rating factors for these concerns.

Table 6.14: Economic Rating Factors for Site Selection

Characteristic	Rating Value
Distance from wastewater source, miles	
0-2	8
2-5	6
5-10	3
>10	1
Elevation difference, feet	
<0	6
0-50	5
50-200	3
>200	1
Land cost and management	
No land purchase, farmer operated	5
Land purchased, farmer operated	3
Land purchased, city or industry operated	1

Procedure for Forested SR Systems: A procedure has been developed for forested SR systems that incorporates climatic, soil, geologic, hydrologic and vegetation factors (11). The procedure involves the use of rating values for subsurface factors (Table 6-15), soils (Table 6-16), and surface factors (Table 6-17).

Table 6.15: Subsurface Factors for Forested SR (11)

Characteristics	Rating Value*
Depth to groundwater on barrier, feet	
<4	0
4-10	4
>10	6
Depth to bedrock, feet	
<5	0
5-10	4
>10	6
Type of bedrock	
Shale	2
Sandstone	4
Granite-gniess	6
Exposed bedrock, percent of total area	
>33	0
10-33	2
1-10	4
none	6

*0-9, site not feasible; 10-13, poor; 14-19, good; and 20-24, excellent.

Table 6.16: Soil Factors for Forested SR (11)

Characteristics	Rating Value*
Infiltration rate, inches/hour	
<2	2
2-6	4
>6	6
Hydraulic conductivity, inches/hour	
>6	2
<2	4
2-6	6
CEC, meq/100 g	
<10	1
10-15	2
>15	3
Shrink-swell potential (SCS)	
High	1
Low	2
Moderate	3
Erosion classification (SCS)	
Severly eroded	1
Eroded	2
Not eroded	3

*5-11, poor; 12-16, good; and 17-21, excellent.

Table 6.17: Surface Factors for Forested SR (11)

Characteristics	Rating Value*
Dominant vegetation	
Pine	2
Hardwood or mixed	3
Vegetation age, years	
Pine	
>30	3
20-30	3
<20	4
Hardwood	
>50	1
30-50	2
<50	3
Mixed pine/hardwood	
>40	1
25-40	2
<25	3
Slope, %	
>35	0
0-1	2
2-6	4
7-35	6
Distance to flowing stream, ft	
50-100	1
100-200	2
>200	3
Adjacent land use	
High density residential/urban	1
Low density residential/urban	2
Industrial	2
Undeveloped	3

*3-4, not feasible; 5-9, poor; 9-14, good; and 15-19, excellent.

Based on the ratings in these tables, an estimate of the preliminary hydraulic loading can be made using Table 6-18. This procedure was developed for sprinkler irrigation of forested sites in the southeastern United States.

Table 6.18: Composite Evaluation of SR Forested Sites (11)

Evaluation Ratings from Tables 6.15 to 6.17			Hydraulic Loading
Poor	Good	Excellent	(inches per week)
3	0	0	Not feasible
2	1	0	<1.0
2	0	1	<1.0
1	2	0	1.0-1.5
1	1	1	1.0-1.5
1	0	2	1.5-2.0
0	3	0	2.0-2.5
0	2	1	2.0-2.5
0	1	2	2.5-3.0
0	0	3	2.5-3.0

REFERENCES

1. Crites, R. W. Land Treatment and Reuse of Food Processing Waste. Proceedings of the Industrial Wastes Symposia. Water Pollution Control Federation. St. Louis, Missouri, 55th Annual Conference. (October 3-8, 1982).

2. Whiting, D. M. Use of Climatic Data in Estimating Storage Days for Soil Treatment Systems. U.S. Environmental Protection Agency. EPA - 600/2-76-250 (November 1976).

3. Moser, M. A., A Method for Preliminary Evaluation of Soil Series Characteristics to Determine the Potential for Land Treatment Processes, Proceedings of the Symposium on Land Treatment of Wastewater, Hanover, New Hampshire (August 1978).

4. Loehr, R.C., Agricultural Waste Management--Problems, Processes, and Approaches. New York: Academic Press (1974).

5. Thomas, R. E. and Whiting, D. M., Annual and Seasonal Precipitation Probabilities. U.S. Environmental Protection Agency, EPA-600/2-77-182 (August 1977).

6. Dewsnup, R. L. and Jensen, D. W. eds., A Summary Digest of State Water Laws. National Water Commission. Washington, D. C. (May 1973).

7. Large, D. W., Land Application of Wastewater and State Water Law: An Overview (Volume I) U. S. Environmental Protection Agency, EPA-600/2-77-232. (November 1977).

8. Large, D. W., Land Application of Wastewater and State Water Law: State Analyses (Volume II). U. S. Environmental Protection Agency. EPA-600/2-78-175. (August 1978).

9. U.S. EPA Process Design Manual for Land Treatment of Municipal Wastewater. U. S. Environmental Protection Agency, EPA 625/1-81-013 (October 1981).

10. Metcalf & Eddy, Inc., Land Application of Wastewater in the Salinas-Monterey Peninsula Area. U. S. Army Engineer District San Francisco, California (April 1976).

11. Taylor, G. L., Land Treatment Site Evaluation in Southeastern Mountainous Areas, Bulletin of the Association of Engineering Geologists, 18:261-266 (1981).

7

Field Investigation Procedures

The factors described in Chapters 4 and 6 regarding groundwater conditions, soil properties and other site factors influence not only the initial site selection and concept feasibility decisions, but are critical for the final system design. The investigation and testing procedures that are commonly used to obtain these data are described in this chapter.

As with all other engineering projects, the type of test required and the specific procedure are relatively easy to describe. The more difficult decision is deciding on how many tests, and in what locations, are adequate for a particular project. Too little field data may lead to erroneous conclusions while too much will result in unnecessarily high costs with little refinement in the design concept. Experience indicates that where uncertainty exists, it is prudent to adopt a conservative posture relative to data gathering requirements.

Table 7-1 is a flow chart which presents a logical sequence of field testing for a land treatment project. When possible, available data are first used for calculations or decisions that may then necessitate additional field tests.

Guidance on testing for wastewater constituents and soil properties is provided for each land treatment process in Table 7-2. Normally, relatively modest programs of field testing and data analysis will be satisfactory.

Table 7.1: Sequence of Field Testing (17)

→ Typical Order of Testing

	Field Tests			
	Test Pits →	Bore Holes →	Infiltration Rate →	Soil Chemistry
Remarks	Usually, with a backhoe, includes inspection of existing SCS reports, road cuts, etc.	Drilled or augered includes inspection of driller's logs for local wells, water table levels	Match the expected method of application, if possible	Includes review of SCS survey
Information to obtain	Depth of profile, texture, structure, soil layers restricting percolation	Depth to groundwater, depth to impermeable layer(s)	Expected minimum infiltration rate	Specific data relating to crop and soil management, phosphorus and heavy metal retention
Estimates now possible	Need for vertical conductivity testing	Groundwater flow direction	Hydraulic capacity based on soil permeability (subject to drainage restrictions	Crop limitations. Soil amendments. Possible preapplication requirements
Additional field tests	Vertical conductivity (optional)	Horizontal conductivity	–	–
Additional estimates	Refinement of loading rates	Mounding analysis, dispersion, need for drainage	–	Quality of percolate
Number of tests	Depends on size, soil uniformity, needed soil tests, type of system. Typical minimum of 3-5 per site	Depends on system type (more for RI than SR), soil uniformity, site size. Typical minimum of 3 per site	Depends on size of site, uniformity of soil. Typical minimum of 2 per site	Depends on uniformity of soil types, type of test, size of site

Table 7.2: Summary of Field Tests for Land Treatment Processes

	Processes		
Properties	Slow Rate (SR)	Rapid Infiltration (RI)	Overland Flow (OF)
Wastewater constituents	Nitrogen, phosphorus, SAR,* EC,* boron	BOD, SS, nitrogen, phosphorus	BOD, SS, nitrogen, phosphorus
Soil physical properties	Depth of profile	Depth of profile	Depth of profile
	Texture and structure	Texture and structure	Texture and structure
Soil hydraulic properties	Infiltration rate	Infiltration rate	Infiltration rate (optional)
	Subsurface permeability	Subsurface permeability	–
Soil chemical properties	pH, CEC, exchangeable cations (% of CEC), EC,* metals,** phosphorus adsorption (optional)	pH, CEC, phosphorus adsorption	pH, CEC, exchangeable cations (% of CEC)

*May be more significant for arid and semiarid areas.

**Background levels of metals such as cadmium, copper, or zinc in the soil should be determined if food chain crops are planned.

SOIL PROPERTIES

A critical element in site selection and process design is the capability of the site soils to move the design quantities of water in the expected direction at the expected rates. These are important requirements for slow rate (SR) systems and are absolutely critical for rapid infiltration (RI) because of the much higher hydraulic loadings. The physical and chemical soil properties of concern are defined and discussed in detail in Chapter 6.

Physical Characteristics

Site identification and selection as described in Chapter 6 will ordinarily be based on existing field data available from a SCS county soil survey and other sources. The next step involves some physical exploration on the site. This preliminary exploration is of critical importance to subsequent phases of the project. Its two purposes are: (1) verification of existing data and (2) identification of probable, or possible, site limitations; and it should be performed with reasonable care. For example, the presence of wet areas, water-loving plant species, or surficial salt crusts should alert the designer to the need for detailed field studies directed toward the problem of drainage. The presence of rock outcroppings would signify the need for more detailed subsurface investigations than might normally be required. If a stream were located near the site, there would need to be additional study of the surface and near-surface hydrology; nearby wells require details of the groundwater flow, and so on. These points may seem obvious. However, there are examples of systems that have failed because of just such obvious conditions: limitations that were not recognized until after design and construction were complete.

The methods of construction and system operation that will

be used can also be critically important depending on the soil properties encountered and must be considered in the site and concept selection process. The characteristics of the soil profile in the undisturbed state may be completely altered when the design surface is exposed or by inadvertant compaction during construction. Fine textured soils are particularly susceptible to compaction. For example, if the design surface layer contains a significant clay fraction and if that surface is exposed for growth of row crops in a SR system the impact of rainfall and sprinkler droplets may result in sorting of the clay fines and a partial sealing of the surface. Such problems can be managed, but the field investigation must provide sufficient data so that such conditions can be anticipated in the design.

RI Systems: Soil properties, topography, and construction methods are particularly critical for RI systems. A site with a heterogeneous mixture of soil types containing scattered lenses of fine textured soil may be impossible to adequately define with a typical investigation program. If such a site cannot be avoided for RI, a large scale pilot test basin is suggested for definition of site hydraulic characteristics. If the pilot test is successful the test basin, if properly located, can then be included in the full scale system.

Sorting of soil fines due to rainfall or turbulent flooding of the RI basin can result in system failure. At Fresno, CA (1), for example, the groundwater recharge RI site was on very flat terrain composed of permeable sandy soils at the surface. A borrow strip was included around the perimeter of each basin to obtain material for dike construction. As a result the final elevation of this borrow strip was depressed relative to the general basin bottom. These borrow strips have been rendered impermeable during the first 10 yr of operation due to sorting and deposition of soil fines and acumulation of organic matter. The applied liquid in this case was high quality river water not wastewater, and the infiltration capacity could not be restored by disking the soil.

An RI site with undulating topography may require a scattered array of basins to remain in desirable soils or may necessitate major cut and fill operations for a compact site. RI basins should always be constructed in cut section if at all possible. Experience (2) has shown that construction in fill sections with soils having a fine fraction (passing #200 sieve) of more than 5% can result in problems. Clayey sands with fines exceeding 10% by weight should be avoided altogether as fill material for basin infiltration surfaces. Pilot scale test basins are recommended whenever RI systems are to be designed on backfilled soils.

Construction: Construction activity either in cut or fill for RI or SR systems can have a drastic effect on soil permeability if clayey sands are present. Such activity should only be permitted when the soil moisture is on the dry side of "optimum". Inadvertant compaction with the soil on the wet side of optimum moisture content could result in the same bulk density for the soil but an order of magnitude reduction in permeability. If such compaction is limited to the top foot of the surface layer, a final ripping and disking may correct the problem. Compaction of this type on sequential layers of fill may not be correctable.

The importance of soil texture for concept and site selection was described in Chapter 4, based on the USDA soil classes (Figure 4-1).

Other suitable soils classification procedures are also in use. These were developed by the American Association of State Highway Officials, (AASHO) and the U.S. Army Corps of Engineers (USACE).

Table 7-3 summarizes the interpretation of these physical and hydraulic properties.

Table 7.3: Interpretation of Soil Physical and Hydraulic Properties

Depth of soil profile, ft	
<1-2	Suitable for OF*
>2-5	Suitable for SR and OF
5-10	Suitable for all processes
Texture and structure	
Fine texture, poor structure	Suitable for OF
Fine texture, well-structured	Suitable for SR and possibly OF
Coarse texture, well-structured	Suitable for SR and RI
Infiltration rate, inches/hour	
0.2-6	Suitable for SR
>2.0	Suitable for RI
<0.2	Suitable for OF
Subsurface permeability	
Exceeds or equals infiltration rate	Infiltration rate limiting
Less than infiltration	May limit application rate

*Suitable soil depth must be available for shaping of overland flow slopes. Slow rate process using a grass crop may also be suitable.

Chemical Properties

The influence of soil chemical properties on permeability and infiltration were discussed in Chapter 4. The importance of pH and soil minerals on fertility is discussed in a later section of this chapter. Adverse chemical reactions between the wastewater and the soil are not expected for municipal and most industrial effluents. The main concern is usually retention or removal of a particular chemical by the soil system and Chapter 3 should be consulted for those details.

Differences in the chemical characteristics between the applied wastewater and the soil may induce chemical changes. At Muskegon, MI, for example, the initial wastewater applications flushed dissolved iron out of the soil profile, showing up as a reddish turbidity in the drain water. At the Fresno, CA system high quality river water (snow melt) was applied to relatively saline soils (1). This low salinity water dispersed the submicron soil colloids in the upper 12 ft of the soil profile. The colloids are then flocculated as mixing occurs with the more saline groundwater. This turbidity problem has persisted for 10 years but does not affect water quality in down gradient wells.

Soil chemistry data is usually obtained via routine laboratory tests on representative samples obtained from test pits or borings. Table 7-4 summarizes the interpretation of typical soil chemical tests.

Table 7.4: Interpretation of Soil Chemical Tests (17)

Test Result	Interpretation
pH of saturated soil paste	
<4.2	Too acid for most crops to do well
4.2-5.5	Suitable for acid-tolerant crops
5.5-8.4	Suitable for most crops
>8.4	Too alkaline for most crops, indicates a possible sodium problem
CEC, meq/100 g	
1-10	Sandy soils (limited adsorption)
12-20	Silt loam (moderate adsorption)
>20	Clay and organic soils (high adsorption)
Exchangeable cations, % of CEC (desirable range)	
Sodium	5
Calcium	60-70
Potassium	5-10
ESP, % of CEC	
<5	Satisfactory
>10	Reduced permeability in fine-textured soils
>20	Reduced permeability in coarse-textured soils
EC_e, mmhos/cm at 25° of saturation extract	
<2	No salinity problems
2-4	Restricts growth of very salt-sensitive crops
4-8	Restricts growth of many crops
8-16	Restricts growth of all but salt-tolerant crops
>16	Only a few very salt-tolerant crops make satis- yields

Test Pits and Borings

Following an initial field reconnaissance, some subsurface exploration will be needed. In the preliminary stages, this consists of digging pits, usually with a backhoe, at several carefully selected locations. Besides exposing the soil profile for inspection and sampling, the purpose is to identify subsurface features that could develop into site limitations, or that point to potential adverse features. Conditions such as fractured, near-surface rock, hardpan layers, evidence of mottling in the profile, lenses of gravel and other anomalies should be carefully noted. For OF site evaluations, the depth of soil profile evaluation can be the top 3 ft or so. The evaluation should extend to 5 ft for SR and 10 ft or more for RI systems.

Representative samples are obtained from the test pits and analyzed to determine the physical and chemical properties discussed above. It is possible with experience to estimate soil texture in the field with small samples taken directly from the walls of the test pit. To determine the soil texture, moisten a sample of soil about 0.5 to 1 in. in diameter. There should be just enough moisture so that the consistency is like putty. Too much moisture results in a sticky material, which is hard to work. Press and squeeze the sample between the thumb and forefinger. Gradually press the thumb forward to try to form a ribbon from the soil. By using this procedure, the texture of the soil can be easily described with the criteria given in Table 7-5.

If the soil sample ribbons (loam, clay loam, or clay), it may be desirable to determine if sand or silt predominate. If there is a gritty feel and a lack of smooth talc-like feel, then sand very likely predominates. If there is a lack of a gritty feel but a smooth talc-like feel, then silt predominates. If there is not a predominance of either the smooth or gritty feel, then the sample should not be called anything other than a clay, clay loam, or loam. If a sample feels quite smooth with little or no grit in it, and will not form a ribbon, the sample would be called silt loam.

Table 7.5: Textural Properties of Mineral Soils (18)

Soil Class	Feeling and Appearance — Dry Soil	Feeling and Appearance — Moist Soil
Sand	Loose, single grains which feel gritty. Squeezed in the hand, the soil mass falls apart when the pressure is released.	Squeezed in the hand, it forms a cast which crumbles when touched. Does not form a ribbon between thumb and forefinger.
Sandy loam	Aggregates easily crushed; very faint velvety feeling initially but with continued rubbing the gritty feeling of sand soon dominates.	Forms a cast which bears careful handling without breaking. Does not form a ribbon between thumb and forefinger.
Loam	Aggregates are crushed under moderate pressure; clods can be quite firm. When pulverized, loam has velvety feel that becomes gritty with continued rubbing. Casts bear careful handling.	Cast can be handled quite freely without breaking. Very slight tendency to ribbon between thumb and forefinger. Rubbed surface is rough.
Silt loam	Aggregates are firm but may be crushed under moderate pressure. Clods are firm to hard. Smooth, flour-like feel dominates when soil is pulverized.	Cast can be freely handled without breaking. Slight tendency to ribbon between thumb and forefinger. Rubbed surface has a broken or rippled appearance.
Clay loam	Very firm aggregates and hard clods that strongly resist crushing by hand. When pulverized, the soil takes on a somewhat gritty feeling due to the harshness of the very small aggregates which persist.	Cast can bear much handling without breaking. Pinched between the thumb and forefinger, it forms a ribbon whose surface tends to feel slightly gritty when dampened and rubbed. Soil is plastic, sticky and puddles easily.
Clay	Aggregates are hard; clods are extremely hard and strongly resist crushing by hand. When pulverized, it has a grit-like texture due to the harshness of numerous very small aggregates which persist.	Casts can bear considerable handling without breaking. Forms a flexible ribbon between thumb and forefinger and retains its plasticity when elongated. Rubbed surface has a very smooth, satin feeling. Sticky when wet and easily puddled.

Beginning at the top or bottom of the pit sidewall, obvious changes in texture with depth are noted. Boundaries that can be seen are marked. When the textures have been determined for each layer, its depth, thickness, and texture layer are recorded.

Table 7.6: Soil Structure Grades (18)

Grade	Characteristics
Structureless	No observable aggregation
Weak	Poorly formed and difficult to see. Will not retain shape on handling.
Moderate	Evident but not distinct in undisturbed soil. Moderately durable on handling.
Strong	Visually distinct in undisturbed soil. Durable on handling.

Soil structure has a significant influence on the soil's acceptance and transmission of water. Soil structure refers to the aggregation of soil particles into clusters of particles, called peds, that are separated by surfaces of weakness. These surfaces of weakness are often seen as cracks in the soil. These planar pores can greatly modify the influence of soil texture on water movement. Well-structured soils with large voids between peds will transmit water more rapidly than structureless soils of the same texture, particularly if the soil has become dry before the water is added. Fine-textured, massive soils (soils with little structure) have very slow percolation rates.

Soil structure can be examined in the pit with a pick or similar device to expose the natural cleavages and planes of weakness. The color and color patterns in soil are also good indicators of the drainage characteristics of the soil. It is often advantageous to prepare the soil pit so the sun will be shining on the face during the observation period. Natural light will give true color interpretations. Artificial lighting should not be used.

Color may be described by estimating the true color for each horizon or by comparing the soil with the colors in a soil color book. In either case, it is particularly important to note the colors or color patterns.

Seasonally high groundwater tables are preferrably detected by borings made during the wet season of the year for the site. An indication of seasonally high groundwater can be observed by the presence of mottles or discolored soils in the wall of the test pit. Mottling in soils is described by the color of the soil matrix and the color or colors, size, and number of the mottles. Each color may be given a Munsell designation and name. However, it is often sufficient to say the soil is mottled. A classification of mottles used by the USDA is shown in Table 7-7.

Reference 8 includes some color photographs of typical soil mottles and can be used to assist in identification.

Table 7.7: Description of Soil Mottles (18)

Character	Class	Limit
Abundance	Few	2% of exposed face
	Common	2-20% of exposed face
	Many	20% of exposed face
Size	Fine	0.25 inch longest dimension
	Medium	0.25-0.75 inch longest dimension
	Coarse	7-75 inches longest dimension
Contrast	Faint	Recognized only by close observation
	Distinct	Readily seen but not striking
	Prominent	Obvious and striking

All of the data collected in the test pit on texture, structure, color, and presence of water should be recorded in the field. A sample log is shown in Figure 7-1.

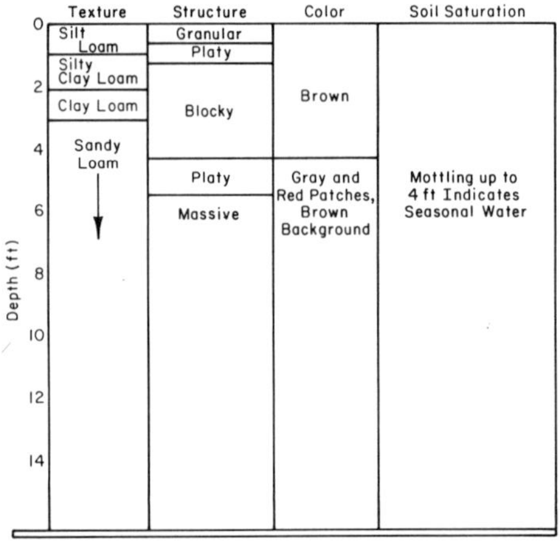

Figure 7-1 Sample log for test pit data.

In some site evaluations, the backhoe pits will not yield sufficient information on the profile. Auger holes or bore holes are frequently used to explore soil deposits below the limits of pit excavation. Augers are useful to relatively shallow depths compared to other boring techniques. Depth limitation for augering varies with soil type and conditions, as well as hole diameter. In unconsolidated materials above water tables, 5 in. diameter holes have been augered beyond 115 ft. Cuttings that are continuously brought to the surface during augering are not suitable for logging the soil materials. Withdrawal of the auger flights for removal of the cuttings near the tip represents an improvement as a logging technique. The best method is to withdraw the flights and obtain a sample with a Shelby tube or split-spoon sampler.

Boring methods, which can be used to probe deeper than augering, include churn drillings, jetting, and rotary drilling. When using any of these methods it is preferable to clean out the hole and secure a sample from the bottom of the hole with a Shelby tube or split-spoon sampler.

GROUNDWATER CONDITIONS

The position, the rate of flow, and the direction of flow of the natural groundwater beneath the site are critical elements in the field investigation. Some key questions to be answered by the investigation are:

1. How deep beneath the surface is the (undisturbed) water table?
2. How does the natural water table depth fluctuate seasonally?
3. How will the groundwater table respond to the proposed wastewater loadings?
4. In what direction and how fast will the mixture of percolate and groundwater move from beneath the area of application? Is there any possibility of transport of contaminants to deeper potable aquifers?
5. What will be the quality of this mixture as it flows away from the site boundaries?
6. If any of the conditions measured or predicted above are found to be unacceptable, what steps can be taken to correct the situation?

Groundwater Depth and Hydrostatic Head

A groundwater table is defined as the contact zone between the free groundwater and the capillary zone. It is the level assumed by the water in a hole extended a short distance below the capillary zone. Groundwater conditions are regular when there is only one groundwater surface and when the hydrostatic pressure increases linearly with depth. Under this condition, the piezometric pressure level is the same as the free groundwater level regardless of the depth below the groundwater table at which it is measured. Referring to Figure 7-2, the water level in the "piezometer" would stand at the same level as the "well" in this condition.

In contrast to a well, a piezometer is a small diameter open pipe driven into the soil such that (theoretically) there can be no leakage around the pipe. As the piezometer is not slotted or perforated, it can respond only to the hydrostatic head at the point where its lower open end is located. The basic difference between water level measurement with a well and hydrostatic head measurement with a piezometer is shown in Figure 7-2.

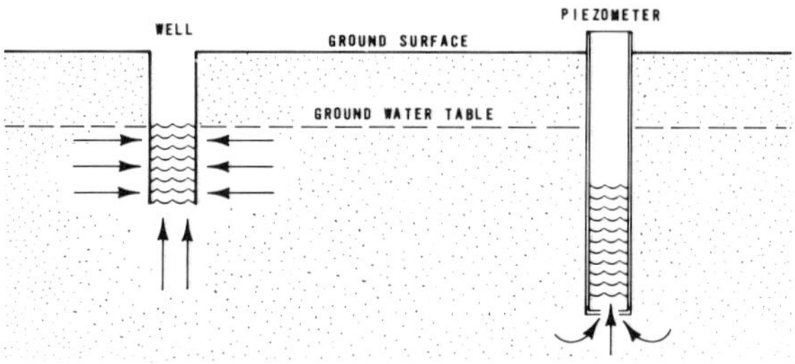

Figure 7-2 Well and piezometer installations, After (17).

Occasionally there may be one or more isolated bodies of water "perched" above the main water table because of lenses of impervious strata that inhibit or even prevent seepage past them to the main body of groundwater below.

Reliable determination of either groundwater levels or pressures requires that the hydrostatic pressures in the bore hole and the surrounding soil be equalized. Attainment of stable levels may require considerable time in impermeable materials. This is called hydrostatic time-lag and may be from hours to days in materials of practical interest.

Two or more piezometers located together, but terminating at different depths, can indicate the presence, direction and magnitude (gradient) of components of vertical flow if such exists. Their use is indicated whenever there is concern about movement of contaminants downward to lower lying aquifers. Figure 7-3 shows several observable patterns with explanations. Reference 6 contains details on the proper installation of wells and piezometers.

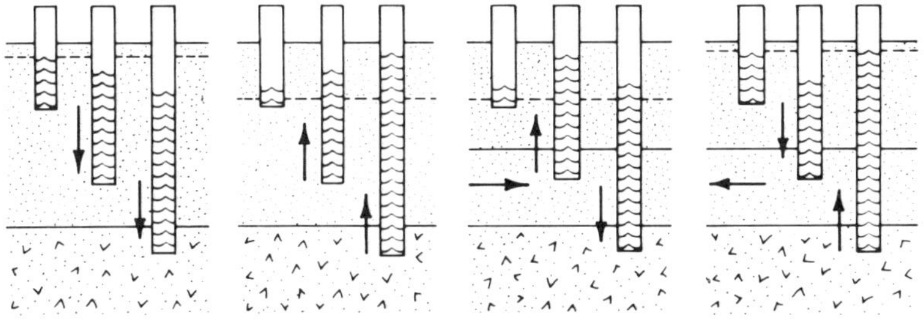

THE PIEZOMETERS INDICATE THAT THE GROUND WATER IS GOING DOWN AND THAT THERE IS SOME NATURAL DRAINAGE.

THE PIEZOMETERS INDICATE A HYDROSTATIC PRESSURE OR THAT THERE IS WATER COMING UP FROM A DEEPER STRATA.

THE PIEZOMETERS INDICATE A HYDROSTATIC PRESSURE IN A STRATUM AND THAT WATER IS BEING FORCED BOTH UP AND DOWN FROM THE STRATUM

THE PIEZOMETERS INDICATE THAT GROUND WATER IS MOVING INTO A STRATUM AND GOING OUT OF THE AREA.

Figure 7-3 Vertical flow direction indicated by piezometers, After (17).

Groundwater Flow

Exact mathematical description of flow in the saturated zones beneath and adjacent to (usually downgradient) land treatment systems is a practical impossibility. However, for the majority of cases the possession of sufficient field data will allow an application of Darcy's equation (see Equation 4-1, and related discussion in Chapter 4) to determine the volume of flow and the mean travel time as well as estimating the mounding that will be created by the wastewater applications. The calculation procedures are presented in detail in Chapter 4. The necessary field data include:

1. Depth to groundwater.
2. Depth to any impermeable barrier.
3. Hydraulic gradient determined from water levels in several observation wells at known distances apart. Establishing the gradient also determines the direction of flow.
4. Specific yield (see Chapter 4).
5. Hydraulic conductivity in the horizontal direction (see Chapter 4 for discussion, a later section in this chapter for test procedures).

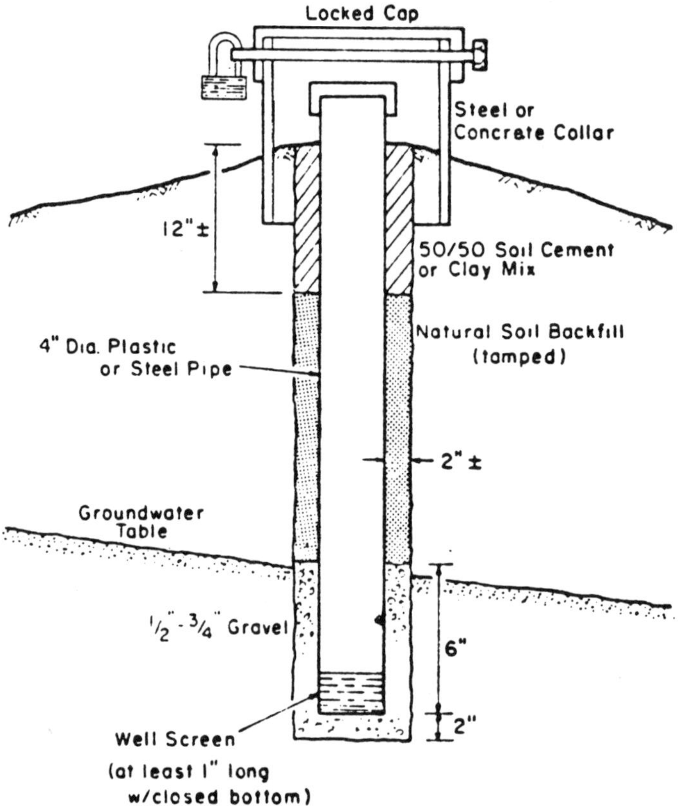

Figure 7-4 Typical shallow monitoring well.

Data for items 1 and 3 can be obtained from periodic water level observations, over a period of months, from simple wells installed on the site. Figure 7-4 illustrates a typical shallow well.

The number and locations required will depend on the size of the project and the complexity of the groundwater system. Typical locations are up gradient of the site, several on the site, and on the down gradient boundary. In general, groundwater levels will tend to reflect the surface contours and flow toward adjacent surface waters. In a complex situation it may be necessary to install a few exploratory wells and then complete the array based on the preliminary data. If properly located, many of these wells can also serve for performance monitoring during system operation. It is necessary to determine the elevation at the top of each well. The depth to water can then be determined with a weighted, chalked tape or other sensing devices. Contours showing equal groundwater elevation can then be interpolated from the well data and plotted on a site map. This in turn allows determination of the hydraulic gradient and the flow direction.

Example 7-1: Groundwater Movement

Conditions: Determine the direction of flow and hydraulic gradient for the situation shown in Figure 7-5. Six monitoring wells were installed on the site. Observed April through October, June had the highest groundwater levels and Figure 7-5 depicts the interpolated contours for that month.

Solution: The flow direction is perpendicular (and down gradient) to the groundwater contours or in this case to the east. The hydraulic gradient is:

$$\frac{\text{difference in groundwater elevation}}{\text{horizontal distance}} = \frac{h}{d} = \frac{(112-106)}{1200} = 0.5\%$$

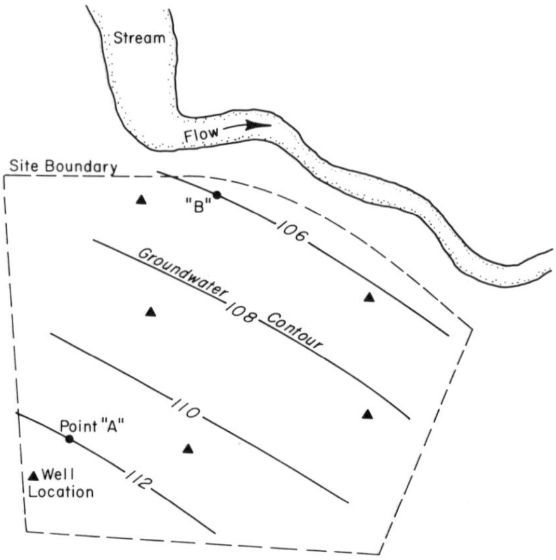

Figure 7-5 Hypothetical groundwater levels for Example 7-1, the horizontal distance from A to B = 1200 ft.

SUBSURFACE PERMEABILITY

The groundwater flow path will be parallel to the hydraulic gradient. In the general situation this is essentially horizontal, except immediately beneath an application zone when mounding occurs. The flow of water will be vertical at the center of the mound and at an angle parallel to the gradient at the edge of the mound. The capability of the soil at the edge of the mound to transmit the applied flow in a lateral direction will control both the vertical development of the mound and its duration in time. The determination of this horizontal conductivity is therefore essential, particularly for RI systems.

Most soils are not homogeneous but rather are at least somewhat stratified, reflecting deposition or consolidation patterns. There are often thin layers or lenses of fine textured material that will impede vertical flow between highly permeable layers of soil. As a result the potential for flow in the horizontal direction is often many times greater than in the vertical direction.

This is illustrated by the ratios in Table 4-3 in Chapter 4. These values are often used for preliminary design calculations. However, in situations with shallow groundwater or where mounding or lateral flow are a significant factor for design it is necessary to measure the horizontal conductivity (K_h) in the field.

Auger Hole Test

The auger hole test is the most common and most useful of the field tests available for determining horizontal hydraulic conductivity. A hole is bored to a certain distance below the water table. The water in the hole is then pumped out. The rate at which the hole refills is a function of the hydraulic conductivity of the soil, and the geometry of the hole. It is possible to calculate the K_h with the measured rate of rise and the other factors defined on Figure 7-6. The general set up for the test is shown in Figure 7-7. The equipment required includes a suitable pump, an auger, a stop watch, and a device for measuring the depth of water in the hole as it rises. In unstable soils a perforated casing or well screens will be necessary to maintain an open hole. The Bureau of Reclamation uses four inch thin wall pipe with 60, 1/8 in. by 1 in. slots per ft of length.

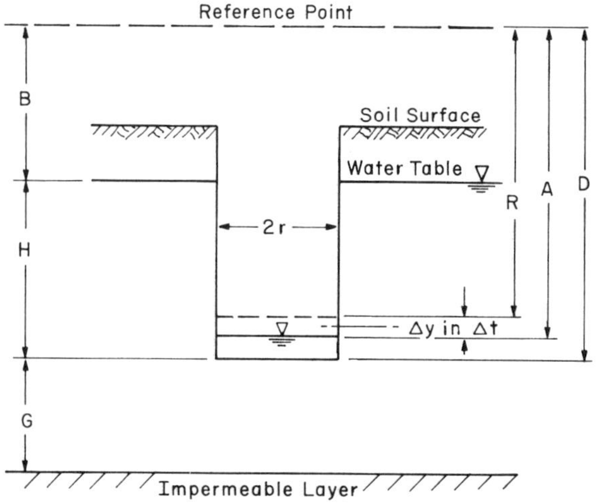

Figure 7-6 Definition sketch for auger hole technique.

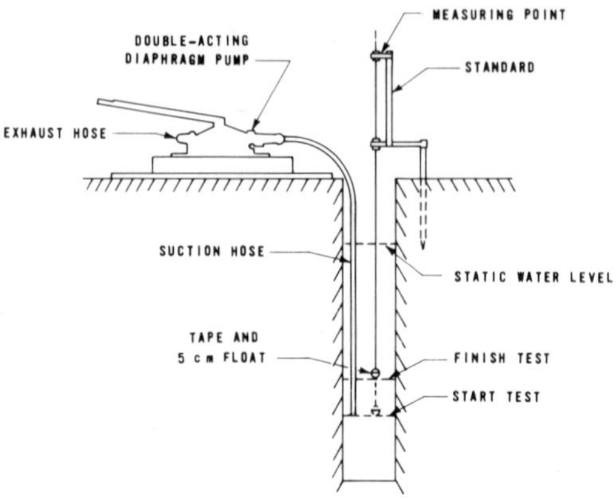

Figure 7-7 Equipment set-up for auger hole test.

The determination of hydraulic conductivity is affected by the location of the barrier or lower impermeable layer. In the case where the barrier is at the bottom of the hole, K_h can be defined as (terms as shown on Figure 7-6):

$$K_h = \frac{15{,}000\ r^2}{(H + 10r)(2-y/H)y} \left(\frac{\Delta y}{\Delta t}\right) \qquad (7\text{-}1)$$

where K_h = horizontal hydraulic conductivity, in/h

r = radius of hole, in.

H = initial depth of water in hole, in.

H = (D-B)

A = depth (from reference point) to water after pumpout, in.

R = depth (from reference point) to water after refill, in.

y = average depth to water in hole during the refill period, in.

y = $(R-B) - 1/2\Delta y$

Δy = raise of water level in the timed interval Δt, in.

Δy = (A-R)

Δt = time required to give Δy, s

The more usual case is when the impermeable layer is some distance below the bottom of the hole; in this case K_h is given by:

$$K_h = \frac{16{,}667\ r^2}{(H + 20r)(2-\frac{y}{H})\ (y)} \quad (\frac{\Delta y}{\Delta t}) \qquad (7\text{-}2)$$

All terms as defined previously.

This equation is only valid when:

$$2\ 1/2 \text{ in.} < 2r < 5\ 1/2 \text{ in.}$$

$$10 \text{ in.} < H < 80 \text{ in.}$$

$$y > 0.2H$$

$$G > H$$

$$y \leqslant 1/4 \quad H-(D-A)$$

Example 7-2: Auger Hole Test

Conditions: Find K_h for the following test conditions: 4 in. diameter hole, 84 in deep, reference point 12 in above ground surface.

Solution: D = 96 in. y = 9 in.

B = 43 in. t = 180s

A = 81 in. R = 72 in.

$$H = (D\text{-}B) = 96\text{-}43 = 53 \text{ in.}$$

$$r = \frac{4}{2} = 2 \text{ in.}$$

$$y = (R\text{-}B) - \frac{1}{2}\Delta y$$

$$= (72\text{-}43) - \frac{1}{2}\ (9)$$

$$= 24.5 \text{ in.}$$

$$G = 12 \text{ ft} = 144 \text{ in.} > H\ (53 \text{ in.})$$

so, use equation 7-2

$$K_h = \frac{(16{,}667)(2)^2}{53 + (20)(2) \quad 2 - \frac{24.5}{53} \quad (24.5)} \quad (\frac{9}{180})$$

$$= 0.96 \text{ in./h}$$

Measurement of horizontal hydraulic conductivity may still be necessary in the absence of a groundwater table. An example might be the presence of fragipan or other hard pan layers at shallow depth. These would restrict vertical flow and might result in unacceptable mounding unless the horizontal conductivity of the overlying material is suitable. The shallow well pump-in test described in Reference 6 can be used in such cases. In effect, it is the reverse of the auger hole test described above. Chapter 15 also describes field tests for small scale systems that can be used to evaluate mounding and lateral flow.

Mixing of Wastewater Percolate with Groundwater

An analysis of the mixing of percolate with native groundwater is needed for SR and RI systems that discharge to groundwater if the quality of this mixture as it flows away from the site boundaries is a concern. The concentration of any constituent in this mixture can be calculated as follows:

$$C_{mix} = \frac{C_P Q_P + C_{gw} Q_{gw}}{Q_P + Q_{gw}} \tag{7-3}$$

where C_{mix} = concentration of constituent in mixture

C_p = concentration of constituent in percolate

Q_p = flow of percolate

C_{gw} = concentration of constituent in groundwater

Q_{gw} = flow of groundwater

The flow of groundwater can be calculated from Darcy's Law (Equation 4-1) if the gradient and horizontal hydraulic conductivity are known. This is not the entire groundwater flow, but only the flow within the mixing depth. Equation 7-3 is only valid if there is complete mixing between the percolate and the native groundwater. This is usually not the case. Mixing in the vertical direction may be substantially less than mixing in the horizontal direction, and density, salinity, and temperature differences between the percolate and groundwater may inhibit mixing and the percolate may in some cases "float" as a plume on top of the groundwater for some distance. The percolation of natural rainfall down gradient of the application site can also serve to dilute the plume.

An alternative approach to estimating the initial dilution is to relate the diameter of the mound developed by the percolate to the diameter of the application area. This ratio has been estimated to be 2.5 to 3.0. This ratio indicates the relative spread of the percolate and can be used to relate the mixing of percolate with groundwater. Thus, an upper limit of 3 for the dilution ratio can be used when groundwater flow is substantially (5 to 10 times) more than the percolate flow. If the groundwater flow is less than 3 times the percolate flow, the actual groundwater flow should be used in Equation 7-3.

INFILTRATION RATE

The infiltration rate of a soil is defined as the rate at which water enters the soil from the surface. When the soil profile is saturated with negligible ponding above the surface, the infiltration rate is equal to the effective saturated conductivity of the soil profile.

Although the measured infiltration rate on a particular site may decrease in time due to surface clogging phenomena, the subsurface vertical permeability at saturation will generally remain constant. Thus, the short-term measurement of infiltration serves reasonably well as an estimate of the long-term saturated vertical permeability if infiltration is measured over a large area.

The value that is required in land treatment design is the long-term acceptance rate of the entire soil surface on the proposed site for the actual wastewater effluent to be applied. The value that can be measured is only a short-term equilibrium acceptance rate for a number of particular areas within the overall site.

There are many potential techniques for measuring infiltration including flooding basin, cylinder infiltrometers, sprinkler infiltrometers and air-entry permeameters. A comparison of these four techniques is presented in Table 7-8. In general, the test area and the volume of water used should be as large as practical. The two main categories of measurement techniques are those involving flooding (ponding over the soil surface) and rainfall simulators (sprinkling infiltrometer). The flooding type of infiltrometer supplies water to the soil without impact, whereas the sprinkler infiltrometer provides an impact similar to that of natural rain. Flooding infiltrometers are easier to operate than sprinkling infiltrometers, but they almost always give higher equilibrium infiltration rates. The sprinkler test is especially useful for agricultural SR operations. As discussed previously, soil sorting and surface sealing can occur with some soils and a sprinkler test will evaluate the possibility. Sprinkler tests are not really needed for grassed or forested sites or where surface application of wastewater is anticipated.

Table 7.8: Comparison of Infiltration Measurement Techniques (17)

Measurement Technique	Water Use per Test (g)	Time per Test (hr)	Equipment Needed	Comments
Flooding basin	600-3,000	4-12	Backhoe or blade	Tensiometers may be used
Cylinder infiltrometer	100-200	1-6	Cylinder or earthen berm	Should use large diameter cylinders (3 ft diameter)
Sprinkler infiltrometer	250-300	1.5-3	Pump, pressure tank sprinkler, cans	For sprinkler applications, soil should be at field capacity before test
Air entry permeameter (AEP)	3	0.5-1	AEP apparatus, standpipe with reservoir	Measures vertical hydraulic conductivity. If used to measure rates of several different soil layers, rate is harmonic mean of conductivities from all soil layers

Since the basic intent of all these tests is to define the saturated vertical hydraulic conductivity of the soil (K_V) and since wastewater will typically be "clean" after a few inches of travel it is usually acceptable to use clean water for these tests. There are exceptions, and the actual wastewater should be used when:

1. High suspended solids or algae are expected in effluents used for RI.
2. Industrial effluents with significantly different pH or ionic composition than the soil and soil water.
3. Effluents that will contain toxic or hazardous materials with potential for reaction with the soil components.

Basin Tests

All infiltration tests should always be run at the actual locations and depths that will be used for the operational system. This is especially important for RI systems. Pilot scale basin tests are strongly recommended. These should be at least 100 ft^2 in area, located in the same soil zone that will be used in the full scale system. Construction of the test basin should be done with the same techniques that will be employed full scale. The test basin should then be operated for several weeks using the same wet and dry cycles that are planned for full scale. Figure 7-8 illustrates a typical small scale pilot test basin.

The number of test basins required will depend on the system size and the uniformity of the soils and topography. One will serve for relatively small systems with uniform soils. In larger systems a separate basin should be used for every major soil type, which may require one basin for every 5-10 acres of total system area. When extremely variable conditions are encountered the test basin should be full sized (1 to 3 acres) to insure reliability. If successful, it can then be incorporated into the operational system.

Figure 7-8 Small scale pilot test basin.

A smaller scale basin type test has been developed by the U.S. Army Corps of Engineers (9). The purpose was to have a reproducible procedure with a larger surface area and zone of influence than existing infiltrometers and permeameters. Figure 7-9 shows the test facility prior to flooding (note the cylinder infiltrometer in the right foreground). The metal ring is aluminum flashing and is 10 ft in diameter. Figures 7-10 and 7-11 provide installation details.

Tensiometers are used in the central part of the test area to insure that saturated conditions prevail during the test period. One should be placed in each soil horizon. In soils lacking well developed horizons a uniform spacing down to about 2 ft will be suitable. Following installation and calibration of the tensiometers, a few preliminary flooding events are executed to achieve saturation. Evidence of saturation is the reduction of tensiometer readings to near zero through the upper soil profile. Then a final flooding event is monitored to derive a cumulative intake versus time curve.

Figure 7-9 U.S. Army Corps of Engineers (USACE) basin test, (from G. Abele).

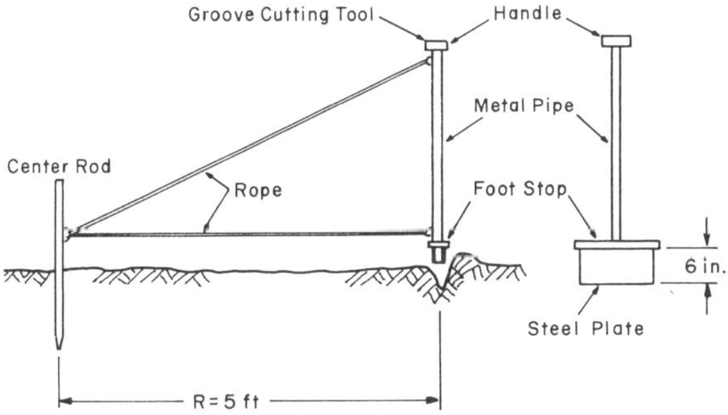

Figure 7-10 Groove preparation for USACE test.

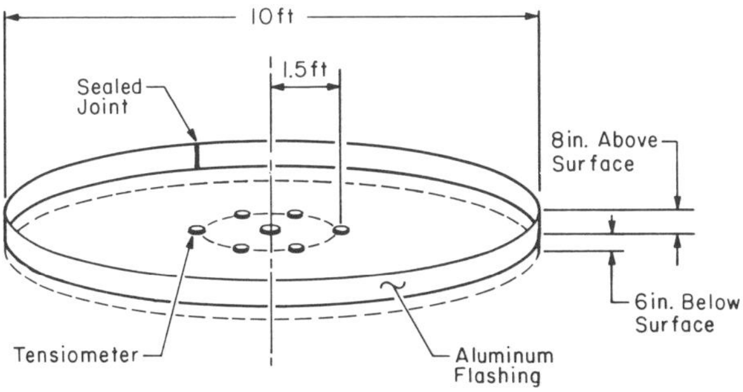

Figure 7-11 Finished installation, USACE test.

Figure 7-12 illustrates typical test results, the "limiting" value of 0.25 in./h was selected for design in this case.

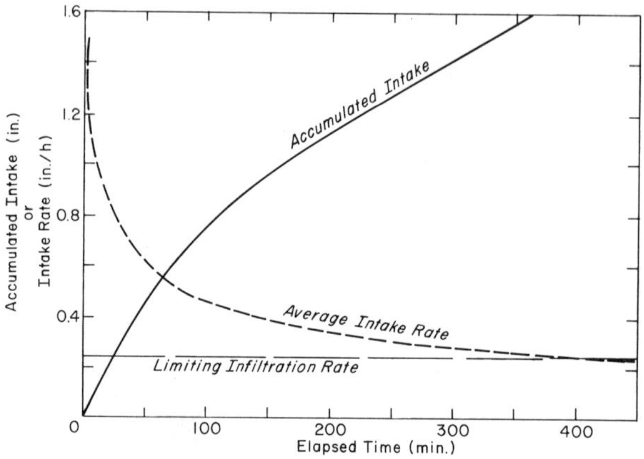

Figure 7-12 Typical test results, USACE infiltration test.

Cylinder Infiltrometers

The equipment setup for a test is shown in Figure 7-13. To run a test, a metal cylinder is carefully driven or pushed into the soil to a depth of about 4 to 6 in. Cylinders from 6 to 14 in diameter have generally been used in practice, with lengths of about 10 to 12 in. Lateral flow is minimized by means of "buffer zone" surrounding the central ring. The buffer zone is commonly provided by another cylinder 16 to 30 in. diameter, driven to a depth of 2 to 4 in. and kept partially full of water during the time of infiltration. This particular mode of making measurements has come to be known as the double-cylinder or double-ring infiltrometer method. Care must be taken to maintain the water levels in the inner and outer cylinders at the same level during the measurements. Alternately, buffer zones are provided by diking the area around the intake cylinder with low (3 to 4 in.) earthen dikes.

If the cylinder is installed properly and the test carefully performed, the technique should produce data that at least approximate the vertical component of flow. In most soils, as the wetting front advances downward through the profile, the infiltration rate will decrease with time and approach a steady-state value asymptotically. This may require as little as 20 to 30 minutes in some soils and many hours in others.

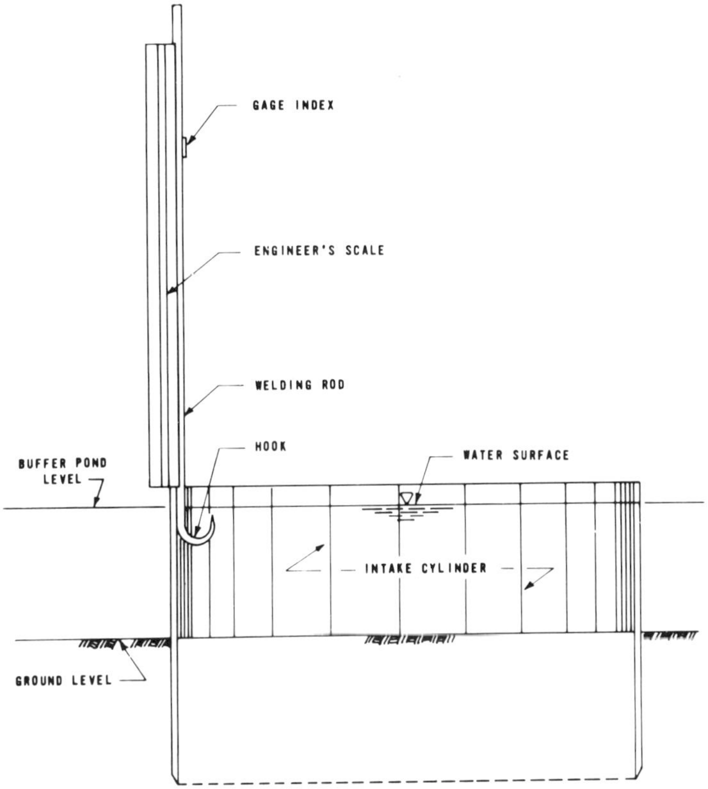

Figure 7-13 Test installation for cylinder infiltrometer.

Test results can be plotted as shown on Figure 7-12 and design values derived. The procedure is relatively simple and quick and uses a small amount of water. The test has been commonly used for some time in agricultural projects and is familiar to most field investigation firms. However, the small size of the test limits the zone of influence. A large number of tests would be required for most situations.

Air Entry Permeameters (AEP)

This device, developed by Dr. Herman Bouwer (10) has been successfully used for the investigation and design of a number of land treatment systems. A sketch of the device is shown on Figure 7-14 and Figure 7-15 illustrates the device in use. The cylinder is steel, about 10 in. in diameter and about 5 in. deep. Operating instructions for the unit are (11):

1. The cylinder is driven into the ground to a depth of 3 to 4 in. (a cylinder driver with sliding weight is used for this purpose).
2. Using a section of 1-in. x 2-in. lumber and a hammer, the soil along the inner perimeter of the cylinder is packed down and against the cylinder wall to insure a good bond between the cylinder and the soil. In loose or cracked soil, compacting around the outside of the cylinder may also be necessary.
3. In case of a bare soil surface, the soil is covered with a 1/2- to 1-in. layer of coarse, clean sand. A disk or similar object is placed on the sand in the center of the cylinder to break the water stream from the supply pipe.

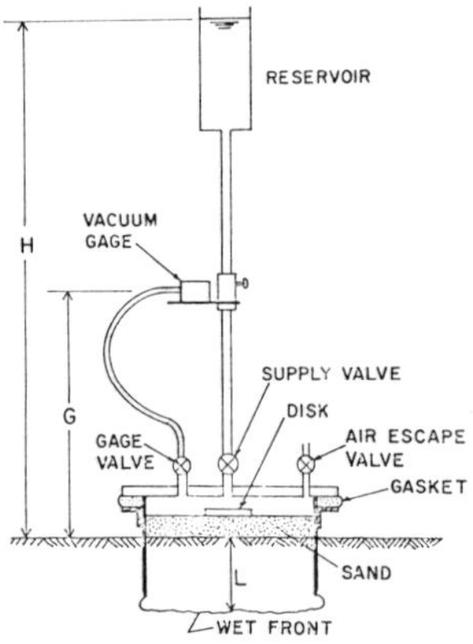

Figure 7-14 Definition sketch for air entry permeameter (AEP) (from H. Bouwer).

Figure 7-15 Air entry permeameter in use (from H. Bouwer).

4. The surface of the foam rubber gasket is cleaned and a thin coat of grease is applied.
5. The lid-assembly with the air valve open and the gauge and supply valves closed is placed on the cylinder. The gauge should be properly primed and air bubbles should not be present in the tubing connecting the gauge to the cylinder. A round bubble-level is placed on the lid to determine the highest point. The lid assembly is then rotated so that the air escape valve is at that highest point.

6. The lid is fastened with four small C-clamps or welder's vice-grip pliers until it rests firmly on the rim of the metal cylinder. Lead weights are placed on the lid to offset the upward hydrostatic force when the supply valve is open.
7. The plastic reservoir at the top of the galvanized pipe is filled with water and the air in the pipe is allowed to escape. The supply valve at the bottom of the galvanized pipe is opened while maintaining the water supply to the plastic reservoir. When the water has driven out the air from inside the cylinder, the air valve is closed.
8. The vacuum gauge is removed from the holder and lifted to about the water level in the plastic reservoir. The gauge valve at the plastic lid is opened, which causes the needle on the gauge to go to zero. Tilting the gauge will then reset the memory pointer to zero. The gauge valve is closed and the gauge is replaced on the gauge holder.
9. Time and water level readings are taken so that the rate of fall of the water level in the reservoir, dH/dt, (just before closing the supply valve) can be calculated.
10. When the depth of the wet front is expected to be at about 4 in. the supply valve is closed. Experience will tell how much or how long water needs to be applied to achieve this depth.
11. The gauge valve is opened. When the gauge indicates approximately atmospheric pressure inside the cylinder, the weights are removed from the plastic lid.
12. When the memory pointer has lost contact with the gauge needle, minimum pressure has occurred. As soon as loss of contact is observed, the memory pointer is read, the gauge valve is closed, and the air escape valve is opened.

The lid assembly is removed and the depth of the wet front is measured. This can be done by pushing a quarter-inch rod into the soil and observing the depth where the penetration resistance is considerably increased. Another way is to quickly remove any remaining water in the cylinder, taking the cylinder out of the soil, and digging with a spade to visually determine the position of the wet front. Dyes and electric-conductivity probes may also offer possibilities for wet-front detection. To facilitate accurate assessment of the depth of the wet front, the soil should not be too wet at the time of the test.

13. Calculate P_a as:

$$P_a = P_{min} + G + L \qquad (7\text{-}4)$$

where P_a = air entry value of soil in in. of water

P_{min} = minimum pressure head in in. water as determined by maximum reading on the vacuum gage

G = height of gage above soil surface, in.

L = depth of wet front, in.

If, for example, the maximum gage reading corresponds to -33 in. water and L + G = 18 in., P_a is calculated as -14 in. water.

14. Calculate the water entry (air exit) value P_w as 0.5 P_a.

15. Calculate the saturated hydraulic conductivity K_s as

$$K_s = 2\,\frac{\frac{dH}{dt}L}{H_t + L - 0.5\,P_a}\;\frac{R_r^2}{R_c^2} \qquad (7\text{-}5)$$

where $\frac{dH}{dt}$ = rate of fall of water level in reservoir just before closing supply valve.

H_t = height above soil surface of water level in reservoir when supply valve is closed.

R_r = radius of plastic reservoir.

R_c = radius of permeameter cylinder.

16. Calculate K at zero soil water pressure head for sorption as 0.5 K_s.

Note: For most agricultural and coarse-textured soils, P_a numerically will be small compared to H_t. Under those conditions, P_a is not important and can be taken as zero (or as some arbitrary small value, for example- 4 in.) in the above equation. This greatly simplifies the equipment and the field procedure, since the vacuum gage and the measurement of minimum pressure inside the cylinder are then not needed.

The AEP test takes less time and less water than cylinder infiltrometers, and the simplicity of the test permits a very large number of repetitions with very small quantities of water. However, the small size of the apparatus limits the zone of influence so the results are only valid for the few inches below the test surface. Several repetitions with depth will be necessary to characterize the soil profile at a particular location. A successful approach is to dig a test pit with a backhoe with one end of the pit inclined to the surface. Benches can then be excavated by hand in the different horizons or at depths of choice and an AEP test run on each "step". The bench should be about 3 ft wide. The other walls of the test pit can then be used for the routine soils investigations. A combination of test basins on the site, supplemented by AEP tests in the remaining areas is recommended as the investigation techniques for most projects.

AGRONOMIC FACTORS

Since SR and OF systems depend on vegetation as an active treatment component it is essential that the field investigation provide sufficient data for reliable design and successful performance of the crop. Important chemical soil properties affecting the vegetation on land treatment systems include: pH, cation exchange capacity, percent base saturation, exchangeable sodium percentage, salinity, plant nutrients, phosphorus and potassium. These factors were discussed in detail in Chapter 3, this section only covers the sampling or testing procedure.

It is recommended that soil samples be collected from each field that will be used. If a given field exceeds 25 acres, individual soil samples should be collected from each soil series within the field. Valid soil sampling procedures are essential. Information can be obtained from university or private soil testing laboratories on proper procedures for obtaining and handling soil samples. The soil analysis should at least determine (1) plant available P and K; (2) soil pH and lime requirement. These tests are routinely performed for most farmers every 2 to 4 years.

In many regions of the United States, a specific soil test is not used to develop N fertilizer needs. Some midwestern states relate N fertilizer applications to soil organic matter, while the nitrate contained in the soil profile is considered in some western states where crops are grown under irrigation.

Samples should be air-dried (at temperatures less than $40^{\circ}C$), ground, and passed through a 2 mm sieve as soon as possible after collection. Chemical analyses are generally performed on air-dried samples, and do not require special preservation for most parameters. However, samples collected for nitrate, ammonia, and pathogen analyses should be refrigerated under field moisture conditions and analyzed as soon as possible.

pH

Soil is prepared for pH determination by making a soil-water paste. When interpreting or using pH data, it is important to know which test method was used because of the influence of the procedure on results. Acid soil conditions (low pH) can be corrected in many cases by the addition of calcium carbonate (lime) to the soil. Alkaline soil conditions (high pH) can be corrected by the addition of acidifying agents. A routine laboratory test procedure is used to estimate the amount of agricultural limestone required to adjust the soil pH. Clover, alfalfa, peas and beans require routine pH adjustment. In the typical case it is not usually economical to apply more than 3 to 4 tons of finely ground limestone per acre at any one time (12). Figure 7-16 shows the amount of ground limestone that would be required to raise the soil pH to 7.0 for typical soils in New York State (13).

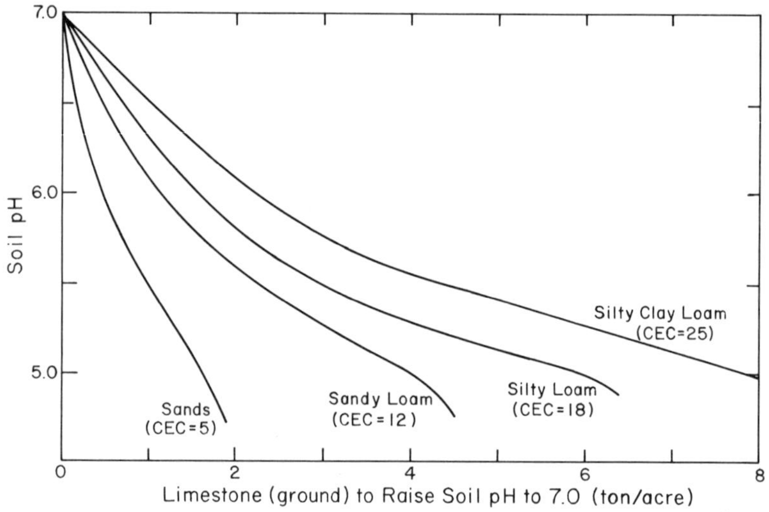

Figure 7-16 Limestone requirements to raise soil pH, After (12).

Plant Available Phosphorus and Potassium

The amount of plant available P is determined by analyzing the amount of P removed from soil by a particular extractant. The extractant used varies in different regions of the United States, but is typically a dilute acid or a bicarbonate solution. Essentially all P taken up by crops is present in insoluble forms in soils rather than being in the soil solution. In all states, it has been determined that there is a relationship between the amount of extractable P in a soil and the amount of P fertilizer needed for various yields of different crops. Such information can be obtained from extension services, universities, etc.

As with P, an extractant is used to determine the plant available K in a soil. Potassium available for plant uptake is present in the soil solution, and is also retained as an exchangeable cation on the cation exchange complex of the soil. The amount of plant available K is then used to determine the K fertilizer rate for the crop grown. Wastewater effluents are usually deficient in K, relative to crop needs in central and eastern parts of the U.S.

Salinity and Sodium

Soils containing excessive exchangeable sodium are termed "sodic" soils. A soil is considered sodic when the percentage of the total CEC occupied by sodium, the exchangeable sodium percentage (ESP), exceeds 15%. These levels of sodium cause clay particles to disperse in the soil because of the chemical nature of the sodium ion. The dispersed clay particles cause low soil permeability, poor soil aeration, and difficulty in seedling emergence. The level of ESP at which these problems are encountered depends on the soil texture. Fine-textured soil may be affected at an ESP above 10%, but coarse-textured soil may not be damaged until the ESP reaches about 20%.

These factors are discussed in detail in Chapter 3. If the field investigation reveals sodic soils or if high salinity wastewater is anticipated, less sensitive crops must be selected as described in Chapter 5.

Test Procedures

References 14, 15, and 16 are the standard sources for soil testing procedures. Most soils laboratories with agronomic capabilities have the capacity for the basic tests discussed in this chapter.

REFERENCES

1. Nightingale, H.I., et al., Leaky Acres Recharge Facility: A Ten Year Evaluation, AWRA Water Resources Bulletin, Vol. 19, 3, p. 429 (June 1983).
2. Reed, S. C. The Use of Clayey Sands for Rapid Infiltration Wastewater Treatment, U.S.A. CRREL IR 805, 55p. (Dec. 1982).
3. Lambe, T. W., Whitman, R.V., Soil Mechanics, J. Wiley, NY, NY, (1969).
4. FHA, Engineering Soil Classification for Residential Development FHA No. 373, U.S. GPO, Wash., D.C. (August 1959).
5. U.S. ACE, The Unified Soil Classification System, Tech Memo 3-357, Vol. 1 (1960 revision) USAE-WES, Vicksburg, MS (1960).
6. U.S, DOI, Drainage Manual, U.S. Department of the Interior, Bureau of Reclamation (1978).
7. Parr, J. F., Marsh, P.B., Kla, J.M., Land Treatment of Hazardous Wastes, Noyes Data Corp., NJ (1983).
8. U.S. EPA, Design Manual-Onsite Wastewater Treatment and Disposal Systems, EPA 625/1-80-012, U.S. EPA Cincinnati, OH (Oct. 1980).
9. Abele, G., McKim, H., Brockett, B., Ingersol, J., Infiltration Characteristics of Soils at Apple Valley, MN. Clarence Cannon Dam, MO., and Deer Creek, OH., Land Treatment Sites. (1980).
10. Bouwer, H., Groundwater Hydrology, McGraw Hill Co., NY (1978).
11. Bouwer, H., Personal Communication (1982).
12. Brady, N.C., The Nature and Properties of Soils, Macmillan Publishing Co., NY (1974).
13. Peech, M., Lime Requirements vs. Soil pH Curves for Soils of New York State, Dept. Agronomy, Cornell University, Ithaca, NY (1961).
14. Black, C.A. ed. Methods of Soil Analysis. American Society of Agronomy, Madison, WI (1965), 1572 p.
15. Walsh, L.M., and Beaton, J.D., eds. Soil Testing and Plant Analysis. Soil Science of America, Madison, WI (1973).

16. Page, A.L. et al, eds. Methods of Soil Analysis - Part 2 - Chemical and Microbiological Properties. 2nd Edition. Soil Science of America, Madison, WI (1983).

17. USEPA, Process Design Manual for Land Treatment of Municipal Wastewater. US Environmental Protection Agency EPA 625/1-81 (October 1981).

18. USEPA, Design Manual - Onsite Wastewater Treatment and Disposal Systems. U.S. Environmental Protection Agency, EPA 625/1-80-012 (October 1980).

8

Preapplication Treatment and Storage

The level of preapplication treatment needed prior to any of the land treatment concepts should be an engineering decision which recognizes the sequence of components as an integrated process. A rational approach would be to start with the final effluent or percolate quality requirements, then determine what contribution the land treatment component can provide, and then adopt a level of preapplication treatment for those materials that will not be removed or reduced to an acceptable concentration by the land. The method of preapplication treatment should then be the simplest and most cost-effective system possible. However, some regulatory agencies still arbitrarily specify both the level and the method of preapplication treatment.

The level of preapplication treatment required should also be based on either the degree of public access to the site or on the end use of the crops. It seems obvious that applying wastewater to irrigate a public playground should require a higher degree of preapplication treatment than applying wastewater to a remote hay field. Table 8-1 summarizes the preapplication treatment guidelines developed by the USEPA. The level of treatment required increases as the degree of public access increases and when the end use of the crop involves direct human consumption. The bacterial standards are based on standard irrigation water requirements in one case and on bathing water standards for the recreational case (1).

Table 8.1: Guidance for Assessing Level of Preapplication Treatment (1)

I. Slow-Rate Systems

A. Primary treatment–acceptable for isolated locations with restricted public access and when limited to crops not for direct human consumption.

B. Biological treatment by lagoons or inplant processes plus control of fecal coliform count to less than 1,000 MPN/100 ml acceptable for controlled agricultural irrigation except for human food crops to be eaten raw.

C. Biological treatment by lagoons or inplant processes with additional BOD or SS control as needed for aesthetics plus disinfection to log mean of 200/100 ml (EPA fecal coliform criteria for bathing waters)–acceptable for application in public access areas such as parks and golf courses.

II. Rapid-Infiltration Systems

A. Primary treatment–acceptable for isolated locations with restricted public access.

B. Biological treatment by lagoons or inplant processes–acceptable for urban locations with controlled public access.

III. Overland-Flow Systems

A. Screening or comminution–acceptable for isolated sites with no public access.

B. Screening or comminution plus aeration to control odors during storage or application–acceptable for urban locations with no public access.

A final reason for preapplication treatment is to reduce the level of suspended solids in the wastewater. High levels of solids can clog sprinklers, valves and other equipment resulting in increased O&M costs. High concentrations of suspended solids or algae may partially clog the infiltration surfaces and reduce the hydraulic capacity of RI systems. Suspended solids concentrations at the levels commonly associated with primary treatment should cause no problems with any of the land treatment concepts. Higher solids levels, typically from industrial sources, may require either some removal or increased maintenance (ie: scarification) of the surface in RI basins, particularly in cool humid climates. The presence of algae is typically of no concern for SR systems. Algae is not effectively removed by OF and can clog the surface in RI systems so the preapplication design must include that consideration.

POLLUTANT REMOVAL IN PONDS

Effluent from any conventional wastewater treatment process can be successfully applied to the land. However, in many cases a pond or lagoon will be the most cost-effective choice. Ponds are often used with land treatment for flow equalization, for emergency storage and when there are seasonal constraints on operation of agricultural systems. In cases where storage is needed it will usually be most cost-effective to combine the treatment and storage functions in a multiple cell lagoon system. Where odor control or high strength wastes are a factor the initial cell may be aerated followed by one or more, deep storage cells. In remote locations an anaerobic primary cell designed for soilds removal and retention may be possible, followed by the storage cells. The treatment occurring in the storage cells will be similar to that in a facultative lagoon. Basic criteria for design and construction of conventional pond systems are available from a number of sources (2)(3).

The pond unit can be specifically designed for the removal of a particular wastewater constituent. More typically, the detention time in the pond component is set by the storage requirements for the system. The removal of various constituents that will occur within that detention time can then be calculated. If additional removal is required the cost-effectiveness of providing more detention time in the pond can be compared to alternative removal processes. The removal of nitrogen in the pond unit is particularly important since, as discussed in earlier chapters, nitrogen is often the Limiting Design Parameter (LDP) for slow rate systems. Any reduction of nitrogen in the pond unit directly impacts on the design of the land treatment component.

BOD and Solids Removal in Lagoons

As indicated in Chapter 3, BOD is usually not the LDP for design of the land treatment component in any of the concepts.

However, many regulatory agencies specify a BOD requirement for the wastewater to be applied so it may be necessary to estimate the removal that will occur in the pond components. There may be a combination of an aerated or anaerobic cell followed by the storage cell(s).

Aerated Cell: The BOD removal that will occur in aerated cells can be estimated with:

$$\frac{C_n}{C_o} = \frac{1}{\left[1 + \frac{k_c t}{n}\right]^n} \quad (8\text{-}1)$$

where C_n = effluent BOD from cell n, mg/L

C_o = influent BOD to system, mg/L

k_c = reaction rate constant (see Table 8-2) at 20°C

t = total hydraulic residence time, d

n = number of cells (typically = 1)

The reaction rate (k_c) is dependent on the liquid temperature:

$$k_{cT} = k_{20}\ \theta^{(T-20)} \quad (8\text{-}2)$$

where k_{cT} = reaction rate at temperature T

k_{20} = reaction rate at 20°C (see Table 8-2)

θ = 1.085

T = pond temperature, °C

Table 8.2: Reaction Rates (k) for Aerated Ponds (3)

Type of Aeration	k at 20°C
Complete mix	2.5
Partial mix	0.276

The temperature of the pond can be estimated with the procedure developed by Eckenfelder (3):

$$T = \frac{T_o + m\,T_A}{1 + m} \quad (8\text{-}3)$$

where T = mean temperature in bulk liquid in pond, °F

T_o = influent temperature, °F

T_A = air temperature, ^oF

m = proportionality factor

= $(1.6)(\frac{t}{d})$

t = time, d

d = depth of pond

The selection of a reaction rate value from Table 8-2 depends on the aeration intensity to be used. The "complete mix" value assumes high intensity aeration sufficient to maintain all solids in suspension. The "partial mix" value assumes that there is sufficient air supplied to satisfy the oxygen demand but that some solids deposition will occur.

The suspended solids from a complete mix cell will equal the concentration in the cell. The suspended solids from a partial mix cell will depend on detention time. For t less than 1 day the solids concentration will be similar to primary effluent.

Facultative Cell: The BOD removal that will occur in a facultative cell can be estimated with:

$$\frac{C_n}{C_o} = e^{-k_p T} \qquad (8\text{-}4)$$

where C_n = effluent BOD, mg/L

C_o = influent BOD, mg/L

k_p = plug flow reaction rate constant dependent on both temperature and on BOD loading rate. Typically assume 0.1 as k_p at 20 C and ignore BOD loading rate variation.

Other procedures are available for estimating BOD removal in facultative ponds (3)(4) but results will be similar to the method presented above. Suspended solids from long detention facultative cells will depend on the algae content but may range from 20 mg/L or less to over 100 mg/L if algae are present.

Anaerobic Cell: An anaerobic cell is more commonly used in the U.S. for high strength, high solids industrial wastewaters. The cell is typically at least 10 ft deep. BOD loadings may be as high as about 450 lb BOD/acre·d, detention times range from 20 to 50 days depending on climate, and a BOD conversion of about 70% is typical. Effluent suspended solids range from 80 to 160 mg/L (5).

A primary anaerobic cell is used at a number of municipal pond systems in the western provinces of Canada (6). These are also designed for solids removal and retention and are typically followed by one or more long detention time facultative cells. Anaerobic cells are usually designed for up to 10 days detention time with depths ranging from 5 to 15 ft. BOD loadings range from about 1,000 lb/acre·d to over 20,000 lb/acre·d. Effluent from these primary cells is comparable to primary effluent. Detectable odors have been noted to at least 1,000 ft around these systems, so a remote location or other odor control is needed. A possibility is to locate the primary anaerobic cell in the center of the system, surrounded by the storage cells and then the land application components.

Nitrogen Removal in Lagoons

The loss of nitrogen from ponds and water bodies has been recognized for some time, but it is only recently that there has been sufficient field data to permit the development of a rational model that could be used for reliable predictions (7). The removal of nitrogen in a pond is dependent on pH, temperature and detention time and under ideal conditions up to 95% removal has been observed. Volatilization of the ammonia faction is believed to be the major pathway responsible for long term permanent losses.

Since nitrogen is often the LDP for land treatment design it is essential to determine the losses that will occur in any preliminary pond units used for treatment or storage. This may influence the basic feasibility of a particular concept, or control the amount of land needed.

The equations presented below can be used for facultative ponds and for storage ponds. The nitrogen losses in short detention time aerated ponds can usually be neglected. The procedure is based on total nitrogen in the system since numerous transformations from one form of N to another are likely during the long detention times.

The basic design Equation is:

$$\frac{N_e}{N_o} = e^{-k_{NT}\left[t + 60.6\ (pH-6.6)\right]} \tag{8-5}$$

where N_e = effluent total N, mg/L

N_o = influent total N, mg/L

k_{NT} = temperature dependent reaction rate

$k_{NT} = k_{N20}\ (1.039)^{(T-20)}$

k_{N20} = reaction rate at 20°C = 0.0064

T = pond temperature, °C

t = detention time, d

pH = median pH in pond during time t

Example 8-1: Nitrogen Removal in Facultative Ponds

Conditions: N_o = 40 mg/L, t = 50 d, pH = 8, depth of pond = 10 ft, T_o = 60°F, T_A = 68°F, determine effluent nitrogen.

Solution:

$$m = (1.6)\left(\frac{t}{d}\right) = (1.6)\left(\frac{50}{10}\right) = 8$$

$$T = \frac{(60) + (8)(68)}{9} = 67^{\circ}F = 19.5^{\circ}C$$

$$k_{NT} = (0.0064)(1.039)^{(19.5-20)}$$

$$= 0.0063$$

$$N_e = (40)\ e^{-0.0063\left[50 + 60.6(8-6.6)\right]}$$

$$= 17 \text{ mg/L}$$

Application of the equation requires information on the wastewater nitrogen concentration, the detention time, pH and temperature conditions to be expected. In a typical case the nitrogen concentration will vary from month to month so actual long term data is desirable for design. A first approximation for typical municipal wastewater can be obtained with:

	Total Nitrogen mg/L
Weak wastewater (BOD ~ 120)	20
Medium wastewater (BOD ~ 220)	35
Strong wastewater (BOD ~ 360)	60

For the first iteration the detention time should be determined based on any BOD removal required or by the storage time needed. If additional N removal is necessary then the cost-effectiveness of providing more detention time can be compared to other alternates (in the previous example calculation an additional 86 days detention time would be needed to reduce the N concentration to less than 10 mg/L for the conditions specified).

The equation is based on plug flow kinetics and is valid when a pond is discharging and the t is then the total detention time in the system. A value of 1/2 t should be used for the filling and storage (non discharge) periods for storage ponds.

The pH is controlled by the algae interactions with the carbonate buffering system in the pond. If possible, pH values should be obtained from an operating pond in the vicinity. The median pH values for four facultative ponds in the U.S. are given in Table 8-3 (8)(9)(10)(11). A rough estimate of the pH to be expected can be obtained with:

$$pH = 7.3\, e^{0.005\,(Alk)} \qquad (8\text{-}6)$$

where pH = median pH in the bulk liquid

Alk = alkalinity of the influent (as $CaCO_3$), mg/L

Table 8.3: Typical pH/Alkalinity Values in Facultative Ponds (17)

Location	pH Annual Median	Alkalinity mg/ℓ Annual Average
Peterborough, NH	7.1	85
Eudora, KS	8.4	284
Kilmichael, MS	8.2	116
Corinne, UT	9.4	557

Phosphorus Removal in Lagoons

Phosphorus removal in the pond may be necessary when stringent P limits are applied to the effluent from OF systems. Alum has been applied, in a 1:1 mole ratio with phosphorus, to pond effluent with the phosphorus then precipitating on the overland flow slope (12). It is also possible to remove phosphorus directly in the pond. A two year study in Ontario, Canada applying alum at 225 mg/L or ferric chloride at 20 mg/L added to the pond influent effectively maintained phosphorus levels in the effluent below 1 mg/L (13). Batch treatment of a lagoon cell prior to discharge has also successfully been demonstrated in Canada (2). Alum, ferric chloride, and lime have all been successfully used. An alum dose of 150 mg/L can produce an effluent phosphorus concentration of less than 1 mg/L and BOD and SS consistently less than 20 mg/L.

Pathogen Removal in Lagoons

The design of systems that include a pond component should evaluate the bacteria and virus reductions that will occur in the pond. In many cases the reductions that will occur in a pond will produce acceptable levels so an extra disinfection step is not required. At Muskegon, MI, for example, the fecal coliforms in the storage pond effluents were consistently below required levels so that chlorination was terminated (14). The effluent in this case is applied to corn, with poultry feed a major use of the harvested corn. Table 8-4 summarizes water quality changes in the storage pond at Muskegon, MI (15) and in a pilot scale pond in Israel (16).

Removal of bacteria and virus in ponds are strongly dependent on temperature and detention time. Figure 8-1 illustrates virus removal in model ponds (17). Similar results were observed at operational facultative ponds in the Southwest, Southeast and North Central U.S. (18). In summer months virus removal exceeded

99% in the first two cells of these systems. The overall removal on a year-round basis exceeded 95%. Removal of fecal coliforms were even higher.

Table 8.4: Changes of Microorganism Concentrations During Storage (14)

	Input (#/100 ml)	Output (#/100 ml)
Muskegon, MI (44) winter-spring storage		
Fecal coliform	10^6	1,000
Haifa, Israel (26) 73 day, winter season		
Total coliform	2.3×10^7	1.84×10^4
Fecal coliform	1.1×10^6	2.4×10^3
Fecal strep.	1.1×10^6	5.0×10^2
Enterovirus	1.1×10^3	0
Haifa, Israel (26) 35 day summer season		
Total coliform	1.4×10^7	2.3×10^4
Fecal coliform	3.5×10^6	2.4×10^4
Fecal strep.	6.0×10^5	3.7×10^3
Enterovirus	200	0

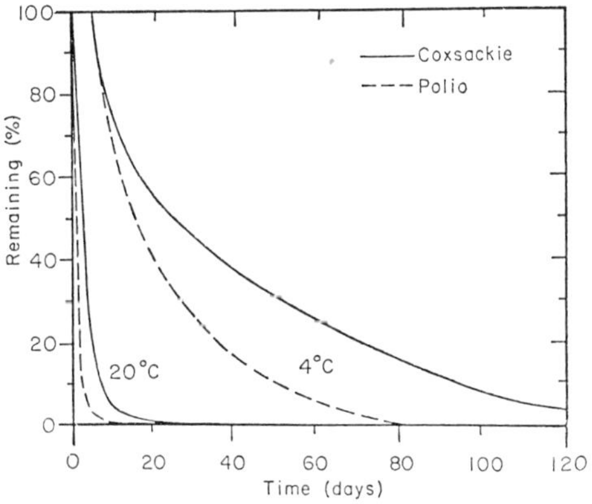

Figure 8-1 Virus removal in ponds, After (17).

Results very similar to Figure 8-1 have been demonstrated for fecal coliforms in facultative ponds in Utah (19). An equation was developed, based on Chick's Law which describes the die-off of fecal coliforms in a pond system as a function of time and temperature:

$$t = \frac{\ln \left[FC_i / FC_f \right]}{(k_{FC})\,(\theta^{T-20})} \qquad (8\text{-}7)$$

where t = actual detention time, d

FC_i = influent fecal coliforms, #/100 ml

FC_f = final fecal coliforms, #/100 ml

k_{FC} = rate constant, use = k_{FC} = 0.5 warm months
= 0.03 in cold months

θ = 1.072

T = liquid temperature in pond, °C

Example 8-2: Pathogen Removal in Facultative Ponds

Conditions: 10 ft deep pond, $t = 90$, $To = 65°F$, $T_A = 70°F$, $FC_i = 10{,}000{,}000/100$ ml. Find effluent Fecal Coliforms.

Solution:

$$m = (1.6)\left(\frac{t}{d}\right) = (1.6)\left(\frac{90}{10}\right) = 14.4$$

$$T = \frac{(65) + (14.4)(70)}{15.4} = 69.7°F = 21°C$$

so: $\theta^{(T-20)} = 1.072$, $k_{FC} = 0.5$

$$FC_f = \frac{FC_i}{e^{(t)(k_{FC})(\theta^{T-20})}}$$

$$= \frac{10{,}000{,}000}{9{,}626{,}209}$$

$$= 1.03/100 \text{ ml}$$

Figure 8-2, based on this equation, illustrates the detention time required to reach either irrigation standards or recreation water standards as a function of temperature. The detention time used in the equation is the actual detention time as measured by dye studies. In the ponds used for model development the actual detention time ranged from 25 to 89 percent of the theoretical design detention time due to short circuiting. The geometric mean was 46%. If the actual detention time is not known it is suggested that this factor be applied when using the equation to estimate fecal coliform die-off to insure a conservative prediction.

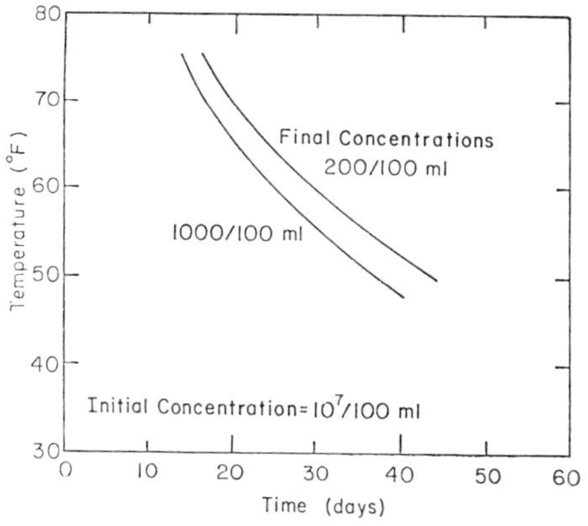

Figure 8-2 Fecal Coliform removal in ponds - detention time needed vs liquid temperature.

Metals and Trace Organic Removal in Lagoons

Removal of metals in the pond component will be comparable to that achieved in primary treatment unless a high intensity complete mix aeration cell is used. In that case removals will be comparable to activated sludge. The removal of trace organics, particularly the volatile type, is very effective in ponds. At the Muskegon system, for example, there were 56 organic compounds in the wastewater entering the ponds. Only 17 of these compounds were present at low concentrations in the effluent leaving the ponds (20).

DESIGN OF STORAGE PONDS

For SR and OF systems, adequate storage must be provided when climatic conditions halt operations or require reduced hydraulic loading rates. Most RI basins are operated year-round, even in areas that experience cold winter weather. Rapid infiltration systems may require cold weather storage during periods when the temperature of the wastewater to be applied is near freezing and the ambient air temperature at the site is below freezing. Generally, the problem occurs only when ponds are used for preapplication treatment. Land treatment systems also may need storage for flow equalization, system backup and reliability, and system management, including crop harvesting (SR and OF) and spreading basin maintenance (RI). Reserve application areas can be used instead of storage for these system management requirements.

During the planning process, Figure 6-2 may be used to obtain a preliminary estimate of storage needs for SR and OF systems. This figure was developed from data collected and analyzed by the National Climatic Center in Asheville, North Carolina. The data were used to develop computer programs that estimate site specific wastewater storage requirements based on climate, which, in turn, were used to plot Figure 6-2. The map is based on the number of freezing days per year corresponding to a 20 year return period. If application rates are reduced during cold weather, additional storage may be required. Should there be a need for more detailed data, the engineer should contact:

Director
National Climatic Center
Federal Building
Asheville, North Carolina 28801
(704) 258-2850

Any communications should refer to computer programs EPA-1, 2, and 3. Each of these programs costs $225 for an initial computer run (January 1981). Table 8-5 summarizes the factors involved with each program and Figure 8-3 illustrates the sections in the U.S. for which each program is applicable.

Table 8.5: Summary of Computer Programs for Determining Storage from Climatic Variables

EPA Program	Applicability	Variables	Remarks
EPA-1	Cold climates	Mean temperature, rainfall, snow depth	Uses freeze index
EPA-2	Wet climates	Rainfall	Storage to avoid surface runoff
EPA-3	Moderate climates	Maximum and minimum temperature, rainfall, snow depth	Variation of EPA-1 for more temperate regions

The storage days are calculated for recurrence intervals of 2, 4, 10, and 20 years.

Storage days required for crop management activities (harvesting, planting, etc.) must be added to the computer estimated storage days due to weather to obtain the total storage days required in each month. The estimated required storage volume is then calculated by multiplying the estimated number of storage days in each month times the average daily flow for the corresponding month.

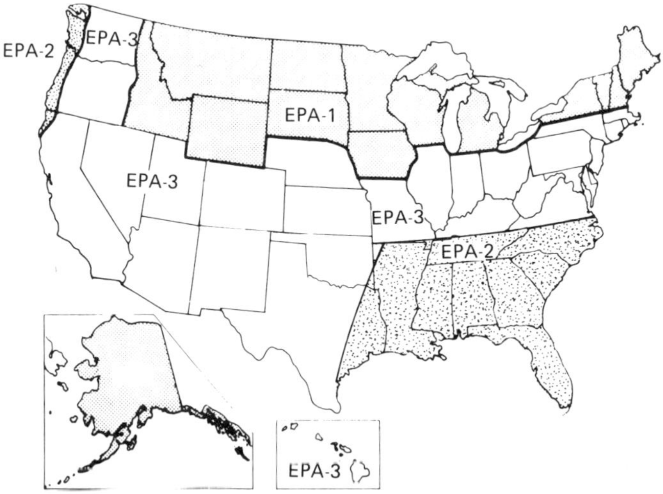

Figure 8-3 Applicable zones in U.S. for application of U.S. EPA storage programs, After (11).

An alternative for preliminary planning of OF and SR systems, is to assume 25°F as the minimum temperature at which a system will successfully operate. Then, the required storage volume is estimated from the average cold weather flow and the number of days in which the mean temperature is less than 25°F.

Storage Calculation Method

The required storage volume should be determined by conducting a monthly water balance which must include the net precipitation or evaporation on the pond. This requires an iterative solution with some assumed initial conditions since the pond area is not known. It is usually convenient to assume a depth for the initial calculation. A water balance for each month can be calculated with the general relationship:

$$S = (P-E) + Q-W-I \tag{8-8}$$

where S = storage volume need for the month,

P = volume of precipitation falling into the pond during the month,

E = volume of water lost by evaporation during the month,

Q = volume of wastewater entering the pond during the month,

W = volume of wastewater leaving the pond during the month,

I = volume of water lost by seepage from the bottom of the pond during the month.

Precipitation (P) and evaporation (E) volumes can be estimated from climatological data. Compile all available monthly and annual data for the closest station to the proposed land treatment site. Using the annual data calculate the 10% chance of exceeding values for both precipitation and evaporation. Apportion these annual values according to the average percent rainfall and evaporation for each month. To convert this data from depth of water (in.) to volume, the surface area of the pond must be known. One method of calculating surface area is to:

1. Estimate the number of storage days required. EPA-1, 2, or 3 programs could be used.

2. Multiply the number of storage days by the average daily design flow to obtain a volume of storage.

3. Assume a depth of water in the pond and calculate surface area by dividing volume of storage by depth. A depth of 10 ft is usually reasonable for a first estimate.

It may be necessary to adjust the surface area once the actual storage volume is determined. However, adjusting surface area requires a recalculation of precipitation and evaporation volumes. The other alternative is to change the depth. The latter method is preferred because P and E volumes then remain unchanged.

The volume of wastewater entering the pond per month (Q) can be calculated by multiplying the average daily design flow by the number of days in the month.

The volume of wastewater leaving the pond (W) can be calculated by multiplying the depth of wastewater applied by the field area. Depth of wastewater applied can be determined from the planned irrigation schedule. Field area (in acres) can be determined from the relationship:

$$F = \frac{Q_a + P_a - E_a}{L_w} \qquad (8\text{-}9)$$

where F = field area, acres

Q_a = volume of wastewater entering the pond, acre-ft/yr,

P_a = volume of precipitation falling into the pond, acre-ft/yr

E_a = volume of evaporation from the pond, acre-ft/yr

L_w = wastewater loading on application site, ft/yr

The volume of water lost by seepage (I) is difficult to estimate over the design life of the system. State standards for allowable seepage rates from ponds vary from 0.062 to 0.25 in/d. These standards are becoming more stringent and essentially impervious linings may be required in the future. Therefore, for a conservative design, assume seepage losses to be negligible.

A typical water balance is shown for assumed conditions in Table 8-6. As shown in the final column of the table the maximum storage required is 5,274 acre-ft during April. If the surface area is maintained at the assumed 430 acres the depth of the pond would have to be 12.3 ft. If the assumed 10 ft depth is retained, then the actual surface will be bigger than 430 acres and another iteration of the calculations will be needed to account for the additional precipitation or evaporation.

Table 8.6: Monthly Water Balance on a Storage Pond

Month	Precip* -Evap. (inches)	P-E Precip. -Evap. (acre-ft)**	Q Wastewater Into Pond*** (acre-ft)	Wastewater Out of Pond† (inches)	Wastewater Out of Pond†† (acre-ft)	S Storage (acre-ft)	Cumulative Storage (acre-ft)
Nov	4.0	143	920	1	103	960	960
Dec	4.2	151	950	0	0	1,101	2,061
Jan	4.3	154	950	0	0	1,104	3,165
Feb	3.5	125	860	0	0	985	4,150
Mar	4.8	172	950	1	103	1,019	5,169
Apr	3.2	115	920	9	930	105	5,274
May	0.7	25	950	15	1,550	-575	4,699
Jun	-1.3	-47	920	20	2,067	-1,194	3,505
Jul	-1.6	-57	950	24	2,480	-1,587	1,918
Aug	-0.3	-11	950	20	2,067	-1,128	790
Sep	0.9	32	920	16	1,653	-701	89
Oct	2.7	98	950	11	1,137	-89	0
Annual	25.1	900	11,190	117	12,090	–	–

*From climatic data for area. Assumes evaporation equal to evapotranspiration.
**Based on a storage pond area of 430 acres. (Estimated 140 days of storage @ 10 mgd and a depth of 10 ft.)
***Based on an average daily design flow of 10 mgd.
†Determined from operational schedule of land application site.
††Based on a land treatment area of 1,240 acres.

Storage for Overland Flow

Storage facilities may be required at an OF system for any of the following reasons:

1. Storage of water during the winter due to reduced hydraulic loading rates or system shutdown.
2. Storage of stormwater runoff to meet mass discharge limitations.
3. Equalization of incoming flows to permit constant application rates.

Due to the limited operating experience with OF in different parts of the country, cold weather storage requirements are not well defined. In general, OF systems must be shut down for the winter when effluent quality requirements cannot be met due to cold temperatures even at reduced application rates or when ice begins to form on the slope. The duration of the shutdown period and, consequently, the required storage period will, of course, vary with the local climate and the required effluent quality.

In studies at Hanover, NH, a storage period of 112 days, including acclimation, was estimated to be required when treating primary effluent to BOD and suspended solids limits of 30 mg/L. This estimate was reasonably close to the 130 storage days predicted by the EPA-1 program using $32^{o}F$ mean temperature. For design purposes, the EPA-1 or EPA-3 programs may be used to conservatively estimate winter storage requirements for OF.

In areas of the country below the 40 day storage contour (on Figure 6-2) OF systems generally can be operated year-round. However, winter temperature data at the proposed OF site should be compared with those at existing systems that operate year-round to determine if all year operation is feasible.

Storage is required at those OF systems that may operate year-round but at reduced hydraulic loading rates during the winter. The required storage volume for such systems can be estimated using the following Equation:

$$V_s = (Q_W)(D_W) - (A_s)(L_{WW})(D_{aw})\left(\frac{7.48}{10^6}\right) \qquad (8\text{-}10)$$

where V_s = storage volume, mg

Q_W = average daily flow during winter, mgd

D_W = number of days in winter period

A_s = slope area, ft^2

L_{WW} = hydraulic loading rate during winter, ft/d

D_{aw} = number of operating days in winter period

The duration of the reduced loading period at existing systems generally has been about 90 days.

Stormwater runoff from the overland flow slopes must be considered since OF is a surface discharging system. In many cases the permits may allow direct discharge of stormwater but may have limitations on the mass of certain constituents that may be present. In such cases, stormwater runoff may need to be stored and discharged at a later time when mass discharge limits would not be exceeded. A procedure for estimating storage requirements for stormwater runoff is outlined below.

1. Determine the maximum monthly mass discharge allowed by the permit for each regulated constituent.

2. Determine expected runoff concentrations of regulated constituents under normal operation (no precipitation).

3. Estimate monthly runoff volumes from the system under normal operation by subtracting estimated monthly ET and percolation losses from design hydraulic loading.

4. Estimate the monthly mass discharge under normal operation by multiplying the values from Steps 2 and 3.

5. Calculate the allowable mass discharge of regulated constituents resulting from storm runoff by subtracting the estimated monthly mass discharge in Step 5 from the permit value in Step 1.

6. Assuming that storm runoff contains the same concentration of constituents as runoff during normal operation,

calculate the volume of storm runoff required to produce a mass discharge equal to the value of Step 5.

7. Estimate runoff as a fraction of rainfall for the particular site soil conditions. Consult the local SCS office for guidance.

8. Calculate the total rainfall required to produce a mass discharge equal to the value in Step 5 by dividing the value in Step 6 by the value in Step 7.

9. Determine for each month a probability distribution for rainfall amounts and the probability that the rainfall amount in Step 8 will be exceeded.

10. In consultation with regulatory officials, determine what probability is an acceptable risk before storm runoff storage is required and use this value (P_d) for design.

11. Storage must be provided for those months in which total rainfall probability exceeds the design value (P_d) determined in Step 10.

12. Determine the change in storage volume each month by subtracting the allowable runoff volume in Step 6 from the runoff volume expected from rainfall having an occurrence probability of P_d. In months when the expected storm runoff exceeds the allowable storm runoff, the difference will be added to storage. In months when allowable runoff exceeds expected runoff, water is discharged from storage.

13. Determine cumulative storage at the end of each month by adding the change in storage during one month to the accumulated quantity from the previous month. The computation should begin at the start of the wettest period. Cumulative storage cannot be less than zero.

14. The required storage volume is the largest value of cumulative storage. The storage volume must be adjusted for net gain or loss due to precipitation and evaporation.

If stored storm runoff does not meet the discharge permit concentration limits for regulated constituents, then the stored water must be reapplied to the OF system. The amount of stored storm runoff is expected to be small relative to the total volume of wastewater applied, and therefore, increases in slope area should not be necessary. The additional water volume can be accommodated by increasing the application period as necessary.

From a process control standpoint it is desirable to operate an OF system at a constant application rate and application period. For systems that do not have storage facilities for other reasons, small equalizing basins can be used to even out flow variations that occur in municipal wastewater systems. A storage capacity of 1 day flow should be sufficient to equalize flow in most cases. The surface area of basins should be minimized to reduce intercepted precipitation. However, an additional half day of storage can be considered to hold intercepted precipitation in wet climates.

For systems providing only screening or primary sedimentation as preapplication treatment, aeration should be provided to keep the basin contents mixed and prevent anaerobic odors. The added cost of aeration, in most cases, will be offset by savings resulting from reduced pump sizes and peak power demands. The designer should analyze the cost-effectiveness of this approach for the system in question.

OPERATION OF STORAGE PONDS

The scheduling of inputs or withdrawals for storage ponds will depend on the overall process and the treatment functions expected for the pond unit.

Storage units in a RI system are typically only for emergency conditions and should be used accordingly. These ponds should remain dry during routine operations and then be drained as rapidly as possible after the emergency is resolved. In some cases a separate pond is not provided in RI systems but extra freeboard constructed in one or more of the infiltration basins.

Storage ponds for OF may be bypassed in many cases during the late spring and summer months to avoid performance problems with algae. The pond contents are then gradually blended with the main wastewater stream so that the pond is drawn down to the specified level at the start of the next storage period. In areas with non-continuous algal blooms the pond discharges should be coordinated with periods of low algae concentration.

Operation of storage ponds for SR systems will depend on whether or not any treatment function has been assigned to the pond. If a specified level of nitrogen or fecal coliform removal is expected then the incoming wastewater should continue to flow into the pond and the withdrawals should be sufficient to reach the required pond level at the end of the application season. When these factors are not a concern, or when it is desired to maximize the nitrogen application to the land, the main wastewater stream should bypass the storage and be applied directly. Regular withdrawals over the season can then draw down the pond. Algae in the pond effluent are not a concern for SR systems so special schedules for this purpose are not required.

PHYSICAL DESIGN AND CONSTRUCTION

Most agricultural storage ponds are constructed earth embankments, the design of which conforms to the principles of small dam design. Depending on the magnitude of the project, state regulations may govern the design. In California, for example, any reservoir with embankments higher than 6 ft and a capacity in excess of 50 acre-ft is subject to state regulations on design and construction of dams, and plans must be reviewed and approved by the appropriate agency. Design criteria and information sources are included in the U.S. Bureau of Reclamation publication, Design of Small Dams (21). In many cases, it will be necessary that a competent soils engineer be consulted for proper soils analyses and structural design of foundations and embankments.

In addition to storage volume, the principal design parameters are depth and area. The design depth and area depend on the function of the pond and the topography at the pond site. If the storage pond is to also serve as a facultative pond, then a minimum water depth of at least 1.5 to 3 ft should be maintained in the pond when the stored volume is at a minimum. The area must also be sufficient to meet the BOD pond loading criteria for the local climate or aerators used to reduce area requirements. The maximum depth depends on whether the reservoir is constructed with dikes or embankments on level ground or is constructed by damming a natural water course or ravine. Maximum depths of diked storage ponds typically range from 9 to 18 ft. Other design considerations include wind fetch, and the need for riprap and lining. These aspects of design are covered in standard engineering references and assistance is also available from local SCS offices, and publications (22).

REFERENCES

1. Thomas, R.E., Reed, S.C., EPA Policy on Land Treatment and the Clean Water Act of 1977, Journal WPCF, 52 (3): 452 (1980).

2. U.S. EPA, Design Manual-Municipal Wastewater Stabilization Ponds, EPA-625/1-83-015, U.S. EPA Cincinnati, OH. (Oct. 1983).

3. Middlebrooks, E.J., et. al., Wastewater Stabilization Lagoon Design, Performance and Upgrading, Macmillan Publishing Co., Inc., NY (1982).

4. Eckenfelder, W.W., Industrial Water Pollution Control, McGraw Hill, Inc., NY (1966).

5. Tchobanoglous, G., Wastewater Engineering: Treatment Disposal Reuse, McGraw Hill, Inc., NY (1979).

6. Higo, T.T., A Study of the Operation of Sewage Ponds in the Province of Alberta. Dept. of Public Health, Government of Alberta, (1966).

7. Reed, S.C., Nitrogen Removal in Wastewater Ponds, U.S.A. CRREL Report No. 84-2, U.S.A. CRREL, Hanover, NH (1984).

8. Reynolds, J.H., et al., Performance Evaluation of an Existing Seven Cell Lagoon System. EPA 600/2-77-086, USEPA, Cincinnati, OH, (1977).

9. McKinney, R.E., Performance Evaluation of an Existing Lagoon System at Eudora, KS, EPA-600/2-77-167, U.S. EPA, Cincinnati, OH, (1977).

10. Hill, D.O., Shindala, A., Performance Evaluation of Kilmichael Lagoon, EPA-600/2-77-109. U.S. EPA, Cincinnati, OH (1977).

11. Bowen, S.P., Performance Evaluation of Existing Lagoons, Peterborough, N.H., EPA-600/2-77-085. U.S. EPA, Cincinnati, OH., (1977).

12. Thomas, R.E., Bledsoe, B., Jackson, K., Overland Flow Treatment of Raw Wastewater with Enhanced Phosphorus Removal. EPA-600/2-76-131, (June 1976).

13. Grahm, H.J., Hunsiger, R.B., Phosphorus Reduction from Continuous Overflow Lagoons by Addition of Coagulants to Influent Sewage, Research Report 65, Ontario Ministry of the Environment, Toronto, Ont., (1977).

14. Reed, S.C., Health Aspects of Land Treatment, U.S. EPA CERI, Cincinnati, OH., U.S. GPO 1979-657-093/7086 (1979)

15. U.S. EPA, Is Muskegon County's Solution, Your Solution? U.S. EPA, Region V, Chicago, IL., (1976).

16. Kott, Y., Lagooned Secondary Effluents as Water Source for Extended Agricultural Purposes. Water Research, 12(12): 1101-1106, (1978).

17. Sagik, B.P., The Survival of Human Enteric Viruses in Holding Ponds, Contract Report DAMD 17-75-C-5062, U.S. Army Medical Research and Development Command, Ft. Detrick, MD., (1978).

18. Bausum, H.T., et al., Enteric Virus Removal in Wastewater Lagoon Systems, U.S. EPA HERL, Report IAG 79-0-X0728, Cincinnati, OH., (1980).

19. Bowles, D.S., Middlebrooks, E.J., Reynolds, J.H., Coliform Decay Rates in Waste Stabilization Ponds. Journal WPCF, 51 (1): 87-99 , (1979).

20. U.S. EPA, Preliminary Survey of Toxic Pollutants at the Muskegon Wastewater Management System, U.S. EPA, ORO, Washington, D.C., (1977).

21. U.S. Department of the Interior, Bureau of Reclamation. Design of Small Dams. Second Edition. U.S. Government Printing Office., (1973).

22. USDA, SCS, Ponds-Planning, Design, Construction Agricultural Handbook, No. 590, U.S. GPO, Washington, D.C.,(1983).

9

Transmission and Distribution Systems

Transmission of wastewater from the point of collection to the land treatment site involves either a pumping station and forcemain, or a gravity pipeline. The wastewater at the site must then be applied using either a surface or sprinkler distribution system.

PUMPING STATIONS

Different types of pumping stations are used for transmission, distribution, and tailwater pumping. Transmission pumping of either raw or treated wastewater usually involves a conventional wastewater pumping station.

Distribution pumping of treated wastewater can involve either a conventional wastewater pumping station or structure built into a treatment/storage pond. Tailwater pumping is used with surface distribution systems and may also be used with some sprinkler distribution systems.

Transmission Pumping

Transmission pumping stations can be located within the wastewater collection system or can be located at the preapplication treatment site if the land application site is remote from the preapplication treatment site. Pumps are usually centrifugal nonclog or vertical turbine.

Figure 9-1 Typical wastewater pumping station.

The number of pumps to be installed depends on the magnitude of the flow and the range of flows expected. Typically the pumps should have capacity equal to the maximum expected inflow with at least one pump out of service.

In pumping stations with capacities of 1 mgd or less usually only two pumps are installed. Each pump should be capable of pumping the maximum inflow. Pumps should be selected with head-capacity characteristics that correspond as nearly as possible to the flow and head requirements of the overall system (1).

The horsepower required for pumping can be estimated using Equation 9-1.

$$H_p = \frac{QH}{3960\ e} \tag{9-1}$$

where H_p = horsepower required, hp

Q = flow, gpm

H = total head, ft

3960 = conversion factor

e = pumping system efficiency

Efficiencies range from about 40 to 50% when pumping raw wastewater up to a range of 65% to 80% when pumping primary or secondary effluent.

Distribution Pumping

Distribution pumping stations can be located next to pre-application treatment facilities or they can be built into the dikes of treatment/storage ponds (see Figure 9-2). Depending on the method of distribution the pumps may discharge under pressure. Peak flows depend on the operation plan and the variation in application rates throughout the operating season. For example, if the land application site is to receive wastewater for only 8 h/d, the pumps must be able to discharge at least three times the average daily flow rate (24 h/d ÷ 8 h/d = 3).

Figure 9-2 Distribution pumps in side of storage pond dike.

The basis of the pump design is the total head (static plus friction) and the peak flow requirements. Flow requirements are determined based on the hours of operation per day or per week and the system capacity (see next section). Details of pumping station design are available in standard references (1)(2).

Tailwater Pumping

Most surface distribution systems will produce some runoff which is referred to as tailwater. When partially treated wastewater is applied, tailwater must be contained within the treatment site and reapplied. Thus, a tailwater return system is an integral part of an SR system using surface distribution methods. A typical tailwater return system consists of a sump or reservoir, a pump (s), and return pipeline (see Figure 9-3).

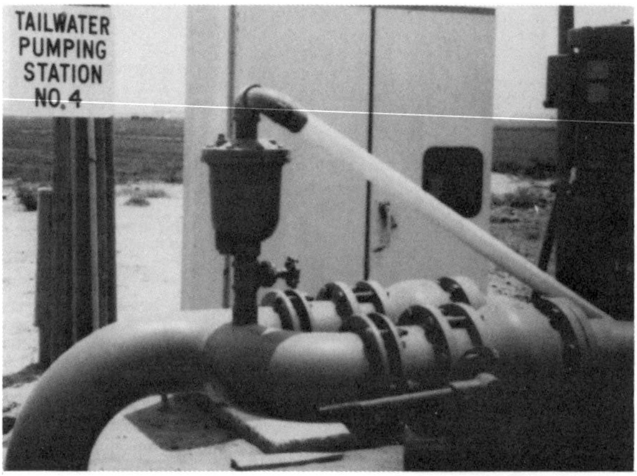

Figure 9-3 Typical tailwater pumping station.

The simplest and most flexible type of system is a storage reservoir system in which all or a portion of the tailwater flow from a given application is stored and either transferred to a main reservoir for later application or reapplied from the tailwater reservoir to other portions of the field. Tailwater return systems should be designed to distribute collected water to all parts of the field, not consistently to the same area. If all the tailwater is stored, pumping can be continuous and can commence at the convenience of the operator. Pumps can be any convenient size, but a minimum capacity of 25% of the distribution system capacity is recommended. If a portion of the tailwater flow is stored, the reservoir capacity can be reduced but pumping must begin during tailwater collection.

Cycling pump systems and continuous pumping systems can be designed to minimize the storage volume requirements, but these systems are much less flexible than storage systems. The designer is directed to reference (3) for design procedures.

The principal design variables for tailwater return systems are the volume of tailwater and the duration of tailwater flow. The expected values of these parameters for a well-operated system depend on the infiltration rate of the soil. Guidelines for estimating tailwater volume, the duration of tailwater flow, and suggested maximum design tailwater volume are presented in Table 9-1.

Table 9.1: Recommended Design Factors for Tailwater Return Systems (3)

. . . Permeability. . . . Class	Rate (in/h)	Texture Range	Maximum Duration of Tailwater Flow % of Application Time	Estimated Tailwater Volume, % of Application Volume	Suggested Maximum Design Tailwater Volume, % of Application Volume
Very slow to slow	0.06-0.2	Clay to clay loam	33	15	30
Slow to moderate	0.02-0.6	Clay loam to silt loam	33	25	50
Moderate to moderately rapid	0.6 6	Silt loams to sandy loams	75	35	70

Runoff of applied wastewater from sites with sprinkler distribution systems should not occur because the design application rate of the sprinkler system is less than the infiltration rate of the soil-vegetation surface. However, some runoff from systems on steep (10 to 30%) hillsides should be anticipated. In these cases, runoff can be temporarily stored behind small check dams located in natural drainage courses. The stored runoff can be reapplied with portable sprinkling equipment.

Forcemains

Forcemains are pressurized pipelines that transmit the wastewater from the pumping station to the application site or storage pond. The considerations in forcemain design are velocity and friction loss. Velocities should be in the range of 3 to 5 ft/s to keep any solids in suspension without developing excessive friction losses. Optimum velocities and pipe sizes depend on the cost of energy and the cost of pipe (see Chapter 13).

Forcemains are usually buried. Pipe materials are usually asbestos cement (AC), ductile iron, or plastic. Under some conditions reinforced concrete pipe (RCP) may also be used.

DISTRIBUTION SYSTEMS

Design of the distribution system involves two steps: (1) selection of the type of distribution system, and (2) detailed design of system components. The two major types of distribution systems are surface and sprinkler systems. Only basic design principles for each type of distribution system are presented in this book, and the designer is referred to several standard agricultural engineering references for further design details (4)(5)(6).

Surface Distribution

With surface distribution systems, water is applied to the ground surface at one end of a field and allowed to spread over the field by gravity. Conditions favoring the selection of a surface distribution system include the following:

1. Capital is not available for the initial investment required for more sophisticated systems.
2. Surface topography of land requires little additional preparation to make uniform grades for surface distribution.

The principal limitations or disadvantages of surface systems include the following:

1. Land leveling costs may be excessive on uneven terrain.
2. Uniform distribution cannot be achieved with highly permeable soils.
3. Runoff control and a return system must be provided when applying wastewater.
4. Periodic maintenance of leveled surfaced is required to maintain uniform grades.

The two general types of surface distribution are the ridge and furrow and the graded border systems. Variations of these two types of methods can be found in standard references (3)(4)(5).

Sprinkler Distribution

Sprinkler distribution uses a rotating nozzle as opposed to spray distribution which refers to a fixed nozzle orifice. Most nozzles used in land treatment systems are of the sprinkler type.

Sprinkler distribution systems simulate rainfall by creating a rotating jet of water that breaks up into small droplets that fall to the field surface. The advantages and disadvantages of sprinkler distribution systems relative to surface distribution systems are summarized in Table 9-2.

Table 9.2: Advantages and Disadvantages of Sprinkler Distribution

Advantages	Disadvantages
Feasible for: porous soils, shallow profiles, rolling terrain, easily eroded soils, small flows, frequent applications	High capital and energy costs
	Traffic problems in clay soils
Positive control of all water	Wind influences distribution and draft
Minimal interference with cultivation	Nozzle clogging
High system efficiency	
No tailwater	

In this book, sprinkler systems are classified according to their movement during and between applications because this characteristic determines the procedure for design. There are three major categories of sprinkler systems based on movement: (1) solid set, (2) move-stop, and (3) continuous move. A summary of the various types of sprinkler systems under each category is given in Table 9-3 along with respective operating characteristics.

Table 9.3: Sprinkler System Characteristics

Type	Typical Application Rate (in./h)	Labor required per Application (h/acre)	Nozzle Pressure Range (lb/in^2)	Size of Single System (acres)	Maximum Grade (%)
Solid set					
Permanent	0.05-2.0	0.008-0.016	30-100	No limit	40
Portable	0.05-2.0	0.03-0.04	30-60	No limit	40
Move-stop					
Hand-move	0.01-2.0	0.08-0.24	30-60	2-40	20
End tow	0.01-2.0	0.03-0.06	30-60	20-40	5-10
Side roll	0.1-2.0	0.016-0.048	30-60	20-80	5-10
Stationary gun	0.25-2.0	0.03-0.06	50-100	20-40	20
Continuous-move					
Traveling gun	0.25-1.0	0.016-0.048	50-100	40-100	20-30
Center pivot	0.20-1.0	0.008-0.024	15-60	40-160	15-20
Linear move	0.20-1.0	0.008-0.024	15-60	40-320	15-20

Design Considerations

Design parameters that are common to all distribution systems are defined as follows:

Depth of Wastewater Applied: The depth of applied wastewater is determined using the relationship:

$$D = \frac{L_W}{F} \tag{9-2}$$

where D = depth of wastewater applied, in.

L_W = monthly hydraulic loading, in.

F = frequency of applications, applications/mo

Application Frequency: The application frequency is defined as the number of applications per month or per week. The application frequency to use for design is a judgment decision to be made by the designer considering: (1) the objectives of the system, (2) the water needs or tolerance of the crop, (3) the moisture retention properties of the soil, (4) the labor requirements of the distribution system, and (5) the capital cost of the distribution system. Some general guidelines for determining an appropriate application frequency are presented here, but consultation with a local farm adviser is recommended.

Except for the water tolerant forage grasses, most crops, including forest crops, require a drying period between applications to allow aeration of the root zone to achieve optimum growth and nutrient uptake. Thus, more frequent applications are appropriate as the ET rate and the soil permeability increase. In practice, application frequencies range from once every 3 or 4 days for sandy soils to about once every 2 weeks for heavy clay soils. An application frequency of once per week is commonly used.

The operating and capital costs of distribution systems can affect the selection of application frequency. With distribution systems that must be moved between applications (move-stop systems), it is usually desirable to minimize labor and operating costs by minimizing the number of moves and therefore the frequency of application. On the other hand, capital costs of the distribution system are directly related to the flow capacity of the system. Thus, the capital cost may be reduced by increasing the application frequency to reduce system capacity.

Application Rate: Application rate is the rate at which water is applied to the field by the distribution system. In general, the application rate should be matched to the infiltration rate of the soil or vegetated surface to prevent excessive runoff and tailwater return requirements. Specific guidelines relating application rates to infiltration properties are discussed under the different types of distribution systems.

Application Period: The application period is the time necessary to apply the desired depth of water (D). Application periods vary according to the type of distribution system, but, in general, are selected to be convenient to the operator and compatible with regular working hours. For most distribution systems application periods are less than 24 hours.

Application Zone: In most systems, wastewater is not applied to the entire field area during the application period. Rather, the field area is divided into application plots or zones and wastewater is applied to only one zone at a time.

Application is rotated among the zones such that the entire field area receives wastewater within the time interval specified by the application frequency. Application zone area can be computed with the following:

$$A_a = A_w/N_a \tag{9-3}$$

where A_a = application zone area

A_w = field area

N_a = No. of application zones.

The number of application zones is equal to the number of applications that can be made during the time interval between successive applications on the same zone as specified by the application frequency.

For example, if the application period is 11 h, effectively 2 applications can be made each operating day. If the application frequency is once per week and the system is operated 7 d per week, then there are 7 operating days between successive applications on the same zone and the number of application zones is:

$$N_a = (2 \text{ applications/day})(7 \text{ operating days})$$

$$= 14$$

If the field area is 35 acres, then the application zone is:

$$A_a = \frac{35}{14}$$

$$= 2.5 \text{ acres}$$

System Capacity: Whatever type of distribution system is selected, the maximum flow capacity of the system must be determined so that components, such as pipelines and pumping stations, can be properly sized. For systems with a constant application rate throughout the application period, the flow capacity of the system can be computed using the following formula:

$$Q = CA_aD/t_a \qquad (9\text{-}4)$$

where Q = discharge capacity, gpm

C = constant, 453

A_a = application area, acres

D = depth of water applied, in.

t_a = application period, h

SURFACE DISTRIBUTION

Ridge and furrow and graded border distribution are usually associated with slow rate systems. For overland flow, surface application can be used with either gated aluminum pipe or bubbling orifices. For rapid infiltration, the common method of application is basin flooding.

Ridge and Furrow Distribution

The design procedure for ridge and furrow systems is empirical and is based on past experience with good irrigation systems and field evaluation of operating systems. For more detailed design procedures, the designer is referred to references (4) and (5).

The design variables for furrow systems include furrow grade, spacing, length, and stream size (flowrate) (Figure 9-4a). The furrow grade will depend on the site topography. A grade of 2% is the recommended maximum for straight furrows. Furrows can be oriented diagonally across fields to reduce grades. Contour furrows or corrugations can be used with grades in the range of 2 to 10%.

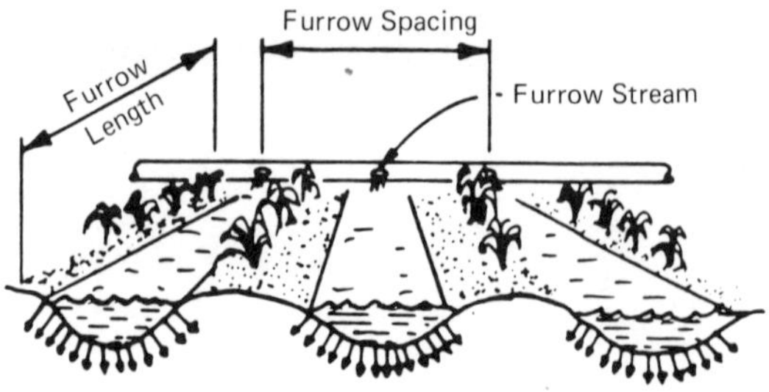

(a) RIDGE AND FURROW

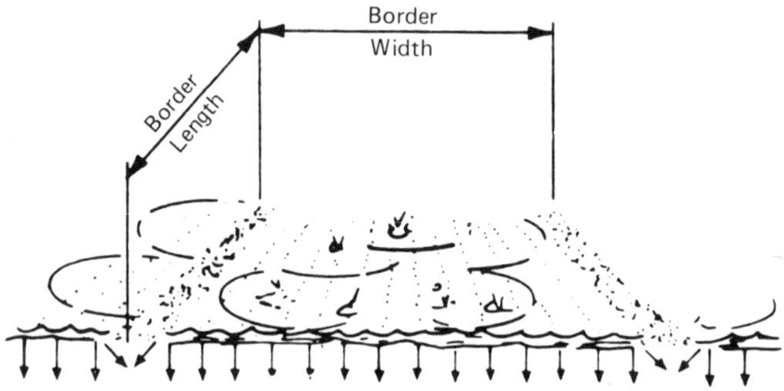

(b) GRADED BORDER

Figure 9-4 Typical surface distribution methods.

The furrow spacing depends on the water intake characteristics of the soil. The principal objective in selecting furrow spacing is to make sure that the lateral movement of the water between adjacent furrows will wet the entire root zone before it percolates beyond the root zone. Suggested furrow spacings based on different soil and subsoil conditions are given in Table 9-4.

The length of the furrow should be as long as will permit reasonable uniformity of application, because labor requirements and capital costs increase as furrows become shorter. Suggested maximum furrow lengths for different grades, soils, and depths of water applied are given in Table 9-5.

Table 9.4: Optimum Furrow Spacing (6)

Soil Condition	Optimum Spacing (inches)
Coarse sands—uniform profile	12
Coarse sands—over compact subsoils	18
Fine sands to sandy loams—uniform	24
Fine sands to sandy loams—over more compact subsoils	30
Medium sandy-silt loam—uniform	36
Medium sandy-silt loam—over more compact subsoils	40
Silty clay loam—uniform	48
Very heavy clay soils—uniform	36

Table 9.5: Suggested Maximum Lengths of Furrows

Furrow Grade (%)	Average Depth of Wastewater Applied* (inches) Clays 3	6	9	12	Loams 2	4	6	8	Sands 2	3	4	5
0.05	1,000	1,300	1,300	1,300	400	900	1,300	1,300	200	300	500	600
0.2	1,200	1,540	1,740	2,030	720	1,200	1,540	1,740	400	600	800	1,000
0.5	1,300	1,640	1,840	2,460	920	1,200	1,540	1,740	400	600	800	1,000
1.0	920	1,300	1,640	1,970	820	980	1,200	1,540	300	500	700	800
2.0	720	890	1,100	1,300	590	820	980	1,100	200	300	500	600

*From Equation 9.2.

The furrow stream size or application rate is expressed as a flowrate per furrow. The optimum stream size is usually determined by trial and adjustment in the field after the system has been installed (5). The most uniform distribution (highest application efficiency) generally can be achieved by starting the application with the largest stream size that can be safely carried in the furrow. Once the stream has reached the end of the furrow, the application rate can be reduced or cut back to reduce the quantity of runoff that must be handled. As a general rule, it is desirable to have the stream size large enough to reach the end of the furrow within one-fifth of the total application period. This practice will result in an application efficiency of greater than 90% for most soils if tailwater is returned.

The application period is the time needed to infiltrate the desired depth of water plus the time required for the stream to advance to the end of the furrow. The time required for infiltration depends on the water intake characteristics of the furrow. There is no standard method for estimating the furrow intake rate. The recommended approach is to determine furrow intake rates and infiltration times by field trials as described in reference (5).

Design of supply pumps and transmission systems should be based on providing the maximum allowable stream size, which is generally limited by erosion considerations when grades are greater than 0.3%. The maximum nonerosive stream size can be estimated from the equation:

$$q_e = C/G \qquad (9\text{-}5)$$

where q_e = maximum unit stream size, gpm.

C = constant, 10

G = grade, %

For grades less than 0.3%, the maximum allowable stream size is governed by the flow capacity of the furrow, estimated as follows:

$$q_c = CF_a \qquad (9\text{-}6)$$

where q_c = furrow flow capacity, gpm

C = constant, 74

F_a = cross-sectional area of furrow, ft^2

For wastewater distribution, pipelines are generally used. If buried pipelines are used to convey water, vertical riser pipes with valves are usually spaced at frequent intervals to release water into temporary ditches equipped with siphon tubes or into hydrants connected to gated surface pipe (Figure 9-5).

Figure 9-5 Typical gated pipe distribution unit.

The spacing of the risers is governed either by the headloss in the gated pipe or by widths of border strips when graded border and furrow methods are alternated on the same field. The valves used in risers are alfalfa valves (mounted on top of the riser) or orchard valves (mounted inside the riser). Valves must be sized to deliver the design flowrate.

Gated surface pipe may be aluminum, plastic, or rubber. Outlets along the pipe are spaced to match furrow spacings. The pipe and hydrants are portable so that they may be moved for each irrigation. The hydrants are mounted on valved risers, which are spaced along the buried pipeline that supplies the wastewater. Operating handles extend through the hydrants to control the alfalfa or orchard valves located in the risers. Control of flow into each furrow is accomplished with slide gates or screw adjustable orifices at each outlet. Slide gates are recommended for use with wastewater. Gated outlet capacities vary with the available head at the gate, the velocity of flow passing the gate, and the gate opening. Gate openings are adjusted in the field to achieve the desired stream size.

Graded Border Distribution

The design variables for graded border distribution are:

1. Grade of the border strip
2. Width of the border strip
3. Length of the border strip
4. Unit stream size

Graded border distribution can be used on grades up to about 7%. Terracing of graded borders can be used for grades up to 20%.

The widths of border strips are often selected for compatibility with farm implements, but they also depend to a certain extent upon grade and soil type, which affect the uniformity of distribution across the strip. A guide for estimating strip widths is presented in Tables 9-6 and 9-7.

Table 9.6: Design Guidelines for Graded Borders for Deep-Rooted Crops (4)

Soil Type and Infiltration Rate (in./h)	Grade (%)	Unit Flow per Foot of Strip Width (gpm)	Average Depth of Water Applied (in)	. . Border Strip (ft) . . Width	Length
Sand					
⩾1.0	0.2-0.4	50-70	4	40-100	200-300
	0.4-0.6	40-50	4	30-40	200-300
	0.6-1.0	25-40	4	20-30	250
Loamy sand					
0.75-1.0	0.2-0.4	30-50	5	40-100	250-500
	0.4-0.6	25-40	5	25-40	250-500
	0.6-1.0	13-25	5	25	250
Sandy loam					
0.5-0.75	0.2-0.4	25-35	6	40-100	300-800
	0.4-0.6	18-30	6	20-40	300-600
	0.6-1.0	9-18	6	20	300
Clay loam					
0.25-0.5	0.2-0.4	13-18	7	40-100	600-1,000
	0.4-0.6	9-13	7	20-40	300-600
	0.6-1.0	5-9	7	20	300
Clay					
0.10-0.25	0.2-0.3	9-18	8	40-100	⩾1,200

Table 9.7: Design Guidelines for Graded Borders for Shallow-Rooted Crops (4)

Soil Profile	Grade (%)	Unit Flow per Foot of Strip Width (gpm)	Average Depth of Water Applied (in.)	. . Border Strip (ft) . . Width	Length
Clay loam, 24" deep over permeable subsoil	0.15-0.6	25-35	2-4	15-60	300-600
	0.6-1.5	18-30	2-4	15-20	300-600
	1.5-4.0	9-18	2-4	15-20	300
Clay, 24" deep over permeable subsoil	0.15-0.6	13-18	4-6	15-60	600-1,000
	0.6-1.5	9-13	4-6	15-20	600-1,000
	1.5-4.0	5-9	4-6	15-20	600
Loam, 6 to 18" deep over hardpan	1.0-4.0	5-20	1-3	15-20	300-1,000

The length of border strips should be as long as practical to minimize capital and operating costs. However, extremely long runs are not practical due to time requirements for patrolling and difficulties in determining stream size adjustments. Lengths in excess of 1,300 ft are not recommended. In general, border strips should not be laid out across two or more soil types with different intake characteristics or water holding capacities, and border strips should not extend across slope grades that differ substantially. The appropriate length for a given site depends on the grade, the allowable stream size, the depth of water applied, the intake characteristics of the soil, and the configuration of the site boundaries. For preliminary design, the length of the border may be estimated using Tables 9-6 and 9-7.

The application rate or unit stream size for graded border irrigation is expressed as a flowrate per unit width of border strip, ft. The stream size must be such that the desired volume of water is applied to the strip in a time equal to, or slightly less than, the time necessary for the water to infiltrate the soil surface. When the desired volume of water has been delivered onto the strip, the stream is turned off. Shutoff normally occurs when the stream has advanced about 75% of the length of the strip. The objective is to have sufficient water remaining on the border after shutoff to apply the desired water depth to the remaining length of border with very little runoff.

Use of a proper stream size is necessary to achieve uniform and efficient application. Too rapid a stream results in inadequate application at the upper end of the strip or in excessive surface runoff at the lower end. If the stream is too small, the lower end of the strip receives inadequate water or the upper end has excessive deep percolation. Actually achieving uniform distribution with minimal runoff requires a good deal of skill and experience on the part of the operator. The range of stream sizes given in Tables 9-6 and 9-7 for various soil and crop conditions may be used for preliminary design. Procedures given in reference (7) may be used to obtain a more accurate estimate of stream size.

The application period necessary to apply the desired depth of water may be determined from the following equation:

$$t_a = LD/Cq \quad (9\text{-}7)$$

where t_a = application period, h

L = border strip length, ft

D = depth of applied water, in.

C = constant, 96.3

q = unit stream size, gpm/ft. of width

The conveyance and application devices used for border distribution are basically the same as described for ridge and furrow distribution. Open ditches with several evenly spaced siphon tubes are often used to supply the required stream size to a border strip. When buried pipe is used for conveyance, vertical risers with valves are usually spaced at intervals equal to the width of the border strip and are located midway in the border strip. With this arrangement, one valve supplies each strip. Water is discharged from the valve directly to the ground surface, as indicated in Figure 9-6, and is distributed across the width of the strip by gravity flow. For border strip widths greater than 30 ft, at least two outlets per strip are necessary to achieve good distribution across the strip. Hydrants and gated pipe can be used with border systems. Use of gated pipe provides much more uniform distribution at the head of border strips and allows the flexibility of easily changing to ridge and furrow distribution if crop changes are desired.

Figure 9-6 Typical discharge valve for border strip application.

Example 9-1: Establish Preliminary Design Criteria for a Graded Border System

Conditions: Deep clay loam soil, finished grade, G: 0.3%, maximum monthly hydraulic loading, L_w: 12 in, application frequency, F : 3 times per month, field area, A_w : 120 acres, crop : pasture.

Solution:

1. Calculate the depth of wastewater to be applied.

$$D = \frac{L_w}{F}$$

$$= \frac{12 \text{ in.}}{3}$$

$$= 4 \text{ in.}$$

2. Select border width and length from Table 9-7 for design conditions for shallow-rooted crops.

Width = 40 ft

Length = 600 ft

3. Select unit flow per width of strip, gpm from Table 9-7.

q = 30 gpm/ft of width

4. Calculate the period of application, t_a, using Equation 9-7.

$$t_a = \frac{LD}{96.3\ q}$$

$$t_a = \frac{(600 \text{ ft})(4)}{(96.3)(30)}$$

$$= 0.83 \text{ h}$$

5. Determine number of applications per day assuming a 12 h/d operating period.

$$\text{Number of applications} = \frac{12 \text{ h/d}}{0.83 \text{ h/application}}$$

$$= 14.5 \text{ applications/d}$$

use 15 applications/d

6. Determine the number of application zones.

Application cycle is 10 day $\left(\frac{30 \text{ d/mo}}{3 \text{ cycles/mo}}\right)$

Application zones = (10d)(15 applications/d)

= 150

7. Calculate the area per zone, A_a.

$$A_a = \frac{A_w}{\text{number of zones}}$$

$$= \frac{120 \text{ acres}}{150 \text{ zones}}$$

$$A_a = 0.8 \text{ acres}$$

8. Determine the number of border strips per application zone.

$$\text{Number of borders} = \frac{A_a}{(L)(W)}$$

$$= \frac{(0.8 \text{ acres})(43,560 \text{ ft}^2/\text{acre})}{(600)(40)}$$

$$= 1.45, \text{ use } 2$$

9. Determine system flow capacity, Q.

$$Q = (2 \text{ borders})(W)(q)$$

$$= (2)(40 \text{ ft})(30 \text{ gpm/ft})$$

$$= 2,400 \text{ gpm}$$

The system must be capable of supplying 2,400 gpm during the maximum month.

Surface Distribution for Overland Flow

Municipal wastewater can be surface applied to overland flow slopes, but industrial wastewater should usually be sprinkler applied. Surface distribution methods include gated aluminum pipe commonly used for agricultural irrigation, and slotted or perforated plastic pipe. Commercially available gated pipe can have gate spaces ranging from 2 to 4 ft and gates can be placed on one or both sides of the pipe. A 2 ft spacing is recommended to provide operating flexibility. Slide gates rather than screw adjustable orifices are recommended for wastewater distribution. Gates can be adjusted manually to achieve reasonably uniform distribution along the pipe. However, the pipe should be operated under low pressure, 2 to 5 lb/in^2, to achieve good uniformity at the application rates recommended in Chapter 11, especially with long pipe lengths. Pipe lengths up to 1,700 ft have been used, but shorter lengths are recommended. For pipe lengths greater than 300 ft, inline valves should be provided along the pipe to allow additional flow control and isolation of pipe segments for separate operation.

Slotted or perforated plastic pipe have fixed openings at intervals ranging from (1 to 4 ft). These systems operate under gravity or very low pressure and the pipe must be level to achieve uniform distribution. Consequently, such methods should be considered only for small systems having relatively short pipe lengths that can be easily leveled. Chapter 11 discusses the advantages and disadvantages of surface and sprinkler systems.

Surface Distribution for Rapid Infiltration

Although sprinklers may be used, wastewater distribution for rapid infiltration is usually by surface spreading. This distribution technique employs gravity flow from piping systems or ditches to flood the application area. To ensure uniform basin application, basin surfaces should be reasonably flat.

Overflow weirs may be used to regulate basin water depth. Water that flows over the weirs is either collected and conveyed to holding ponds for recirculation or distributed to other infiltration basins. If each basin is to receive equal flow, the distribution piping channels should be sized so that hydraulic losses between outlets to basins are insignificant. Design standards for distribution systems and for flow control and measurement techniques are published by the American Society of Agricultural Engineers (ASAE). Outlets used at currently operating systems include valved risers for underground piping systems and turnout gates from distribution ditches.

Basin layout and dimensions are controlled by topography, distribution system hydraulics, and loading rate. The number of basins is also affected by the selected loading cycle. As a minimum, the system should have enough basins so that at least one basin can be loaded at all times, unless storage is provided.

The number of basins also depends on the total area required for infiltration. Optimum basin size can range from 0.5 to 5 acres for small to medium sized systems to 5 to 20 acres for large systems. For a 62 acre system, if the selected loading cycle is 1 day of wastewater application alternated with 10 days of drying, a typical design would include 22 basins of 2.8 acres each. Using 22 basins, 2 basins would be flooded at a time and there would be ample time for basin maintenance before each flooding period.

At many sites, topography makes equal-sized basins impractical. Instead, basin size is limited to what will fit into areas having suitable slope and soil type. Relatively uniform loading rates and loading cycles can be maintained if multiple basins are constructed. However, some sites will require that loading rates or cycles vary with individual basins.

In flat areas, basins should be adjoining and should be square or rectangular to maximize land use. In areas where groundwater mounding is a potential problem, less mounding occurs when long, narrow basins with their length normal to the prevailing groundwater flow are used than when square or round basins are constructed. Basins should be at least 12 in. deeper than the maximum design wastewater depth, in case initial infiltration is slower than expected and for emergencies. Basin walls are normally compacted soil with slopes ranging from 1:1 to 1:2 (vertical distance to horizontal distance). In areas that experience severe winds or heavy rains, basin walls should be planted with grass or covered with riprap to prevent erosion.

If basin maintenance will be conducted from within the basins, entry ramps should be provided. These ramps are formed of compacted soil at grades of 10 to 20% and are from 10 to 12 ft wide. Basin surface area for these ramps and for wall slopes should not be considered as part of the necessary infiltration area.

SPRINKLER DISTRIBUTION

Sprinkler distribution is common to SR systems, is generally used with industrial OF systems, and can be used with RI systems. Forest SR, OF and many agricultural SR systems use solid set (stationary) sprinkler distribution, whereas move-stop and continuous move sprinklers are restricted to SR systems.

For all SR sprinkler systems the design application rate (in/h) should be less than the infiltration rate of the surface soil to avoid surface runoff. For final design, the application rate should be based on field infiltration rates determined from previous experience with similar soils and crops or from direct field measurements.

Solid Set Systems

Solid set sprinkler systems remain in one position during the application season. The system consists of a grid of mainline and lateral pipes covering the field to be irrigated. Impact sprinklers are mounted on riser pipes extending vertically from the laterals. Riser heights are determined by crop heights and spray angle. Sprinklers are spaced at prescribed equal intervals along each lateral pipe, usually 40 to 100 ft. A system is called fully permanent or stationary when all lines and sprinklers are permanently located. Permanent systems usually have buried main and lateral lines to minimize interference with farming operations. Solid set systems are called fully portable when portable surface pipe is used for main and lateral lines. Portable solid set systems can be used in situations where the surface pipe will not interfere with farming operations and when it is desirable to remove the pipe from the field during periods of winter storage. When the mainline is permanently located and the lateral lines are portable surface pipe, the system is called semipermanent or semiportable.

The primary advantages of solid set systems are low labor requirements and maintenance costs, and adaptability to all types of terrain, field shapes, and crops. They are also the most adaptable systems for climate control requirements. The major disadvantages are high installation costs and obstruction of farming equipment by fixed risers.

Application Rate: For solid set systems, the application rate is expressed as a function of the sprinkler discharge capacity, the spacing of the sprinklers along the lateral, and the spacing of the laterals along the main according to the following equation:

$$R = q_s C / S_s S_L \quad (9\text{-}8)$$

where R = application rate, in/h

q_s = sprinkler discharge rate, gpm

C = constant = 96.3

S_s = sprinkler spacing along lateral, ft

S_L = lateral spacing along main, ft

Detailed procedures for sprinkler selection and spacing determination to achieve the desired application rate are given in references (8-10).

Sprinkler Selection and Spacing Determination: Sprinkler selection and spacing determination involves an iterative process. The usual procedure is to select a sprinkler and lateral spacing, then determine the sprinkler discharge capacity required to provide the design application rate at the selected spacing. The required sprinkler discharge capacity may be calculated using Equation 9-8.

Manufacturers' sprinkler performance data are then reviewed to determine the nozzle sizes, operating pressures, and wetted diameters of sprinklers operating at the desired discharge rate. The wetted diameters are then checked with the assumed spacings for conformance with spacing criteria. Recommended spacings are based on a percentage of the wetted diameter and vary with the wind conditions. Recommended spacing criteria are given in Table 9-8.

The sprinkler and nozzle size should be selected to operate within the pressure range recommended by the manufacturer. Operating pressures that are too low cause large drops which are concentrated in a ring a certain distance away from the sprinkler, whereas high pressures result in fine drops which fall near the sprinkler. Sprinklers with low design operating pressures are desirable from an energy conservation standpoint.

Table 9.8: Recommended Spacing of Sprinklers (8)

Average Wind Speed (mph)	Spacing (% of wetted diameter)
(0-7)	40 (between sprinklers) 65 (between laterals)
(7-10)	40 (between sprinklers) 60 (between laterals)
(>10)	30 (between sprinklers) 50 (between laterals)

Lateral Design: Lateral design consists of selecting lateral sizes to deliver the total flow requirement of the lateral with friction losses limited to a predetermined amount. A general practice is to limit all hydraulic losses (static and dynamic) in a lateral to 20% of the operating pressure of the sprinklers. This will result in sprinkler discharge variations of about 10% along the lateral. Since flow is being discharged from a number of sprinklers, the effect of multiple outlets on friction loss in the lateral must be considered. A simplified approach is to multiply the friction loss in the entire lateral at full flow (discharge at the distal end) by a factor based on the number of outlets. The factors for selected numbers of outlets are presented in Table 9-9. For long lateral lines, capital costs may be reduced by using two or more lateral sizes that will satisfy the headloss requirements.

The following guidelines should be used when laying out lateral lines:

1. Where possible, run the lateral lines across the predominant land slope and provide equal lateral lengths on both sides of the mainline.
2. Avoid running laterals uphill where possible. If this cannot be avoided, the lateral length must be

shortened to allow for the loss in static head.

3. Lateral lines may be run down slopes from a mainline on a ridge, provided the slope is relatively uniform and not too steep. With this arrangement, static head is gained with distance downhill, allowing longer or smaller lateral lines to be used compared to level ground systems.

4. Lateral lines should run as nearly as possible at right angles to the prevailing wind direction. This arrangement allows the sprinklers rather than laterals to be spaced more closely together to account for wind distortion and reduces the amount of pipe required.

Table 9.9: Pipe Friction Loss Factors to Obtain Actual Loss in a Line with Multiple Outlets (6)

Number of Outlets	Value of F
1	1.000
2	0.634
3	0.528
4	0.480
5	0.451
6	0.433
7	0.419
8	0.410
9	0.402
10	0.396
15	0.379
20	0.370
25	0.365
30	0.362
40	0.357
50	0.355
100	0.350

<u>Example 9-2: Establish Preliminary Design Criteria for Solid Set Sprinkler System</u>

<u>Conditions</u>: Infiltration rate: 0.6 in/h, depth of wastewater applied, D : 2 in., crop : forage grass, application zone area, A_a : 10 acres, average wind speed : 5 mph.

<u>Solution</u>:

1. Determine design application rate, R. Assume an 8 h application period.

$$R = \frac{D}{t_a}$$

$$= \frac{2 \text{ in.}}{8 \text{ h}}$$

$$= 0.25 \text{ in./h} \ (\quad 0.6 \text{ in./h})$$

2. Select sprinkler and lateral spacings.

use $S_s = 60$ ft

$S_L = 60$ ft

3. Calculate required sprinkler discharge using Equation 9-8.

$$q_s = \frac{R\, S_s S_L}{96.3}$$

$$q_s = \frac{(0.25)(60)(60)}{96.3}$$

$$= 9.3 \text{ gpm}$$

4. Select sprinkler nozzle size, pressure, and wetted diameter to provide necessary discharge.

Use a 7/32 in. nozzle at 50 lb/in.^{2} pressure.

Wetted diameter = 125 ft

5. Check selected spacing against criteria in Table 9-8 for the average wind speed.

Sprinkler spacing, $S_s = \frac{60}{125}$

$= 48\% \quad 40\%$ (too large)

Lateral spacing, $S_L = \frac{60}{125}$

$= 48\% \quad 65\%$ (OK)

6. Change sprinkler spacing to 50 ft (OK at 40%), and lateral spacing to 80 ft (OK at 64%). Recalculate $q_s = 10.4$ gpm The same nozzle is satisfactory if the pressure is increased to 55 lb/in^2. Wetted diameter is 127 ft.

7. Determine system flow capacity, Q.

$$Q = A_aR$$

$$= (10 \text{ acres})(0.25 \text{ in/h})(27{,}154 \text{ gal/acre·in})\left(\frac{1\text{hr}}{60\text{ min}}\right)$$

$$= 1{,}131 \text{ gpm}$$

Solid Set Forest Systems

Solid-set irrigation systems are the most commonly used systems in forests. Buried systems are less susceptible to damage from ice and snow and do not interfere with forest management activities (thinning, harvesting, and regeneration). Solid set sprinkler systems for forest crops have some special design requirements. Spacing of sprinkler heads must be closer and operating pressures lower in forests than other vegetation systems because of the interference from tree trunks and leaves and possible damage to bark. A 60 ft spacing between sprinklers and an 80 ft spacing between laterals has proven to be an acceptable spacing for forested areas. This spacing, with sprinkler overlap, provides good wastewater distribution at a reasonable cost. Operating pressures at the nozzle should not exceed 55 $lb/in.^2$, although pressures up to 85 $lb/in.^2$ may be used with mature or thickbarked hardwood species. The sprinkler risers should be high enough to raise the sprinkler above most of the understory vegetation, but generally not exceeding 5 ft. Low-trajectory sprinklers should be used so that water is not thrown into the tree canopies, particularly in the winter when ice buildup on pines and other evergreen trees can cause the trees to be broken or uprooted.

A number of different methods of applying wastewater during subfreezing temperatures in the winter have been attempted. These range from various modifications of rotating and nonrotating sprinklers to furrow and subterranean applications. General practice is to use low-trajectory, single nozzle impact-type sprinklers, or low-trajectory, double nozzle hydraulic driven sprinklers. A spray nozzle used at West Dover, Vermont, is shown in Figure 9-7.

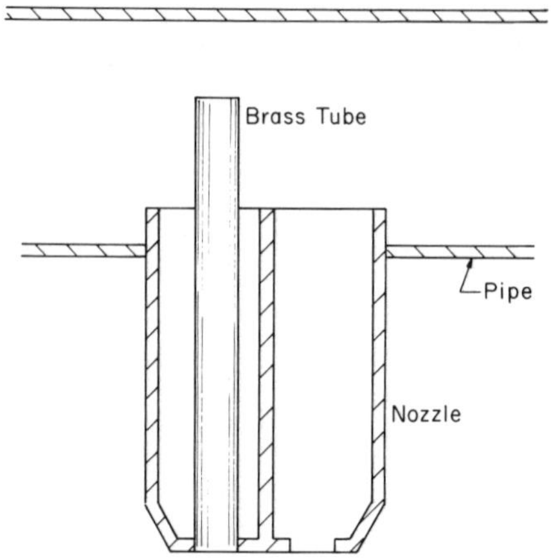

Figure 9-7 Nozzle adaptation for winter spraying. The brass tube drains quickly when the pipe flow is stopped. The other orifice drains more slowly and may freeze. Discharge will start immediately out of brass tube at start of next cycle. Heat from this fluid then melts any ice in other side.

Installation of a buried solid-set irrigation system in existing forests must be done with care to avoid excessive damage to the trees or soil. Alternatively, solid-set systems can be placed on the surface if adequate line drainage is provided (see Figure 9-8). For buried systems, sufficient vegetation must be removed during construction to ensure ease of installation while minimizing site disturbance so that site productivity is not decreased or erosion hazard increased. A 10 ft wide path cleared for each lateral meets these objectives. Following construction, the disturbed area must be mulched or seeded to restore infiltration and prevent erosion. During operation of the land treatment system, a 5 ft radius should be kept clear around each sprinkler. This practice allows better distribution and more convenient observation of sprinkler operation. Spray distribution patterns will still not meet agricultural standards, but this is not as important in forests because the roots are quite extensive.

Figure 9-8 Forest solid set system.

Solid Set Overland Flow Systems

Sprinkler distribution systems recommended for OF systems are discussed in Chapter 11. High pressure, 50 to 80 lb/in^2, impact sprinklers have been used successfully with food processing wastewaters containing suspended solids concentrations >500 mg/L. The position of the impact sprinkler on the slope is also discussed in Chapter 11.

The spacing of the sprinkler along the slope depends on the design application rate and must be determined in conjunction with the sprinkler discharge capacity and the spray diameter. The relationship between OF application rate and sprinkler spacing and discharge capacity is given by the following Equation:

$$R = \frac{q}{S_s} \qquad (9\text{-}9)$$

where R = OF application rate, gpm/ft of slope width

q = sprinkler discharge rate, gpm

S_s = sprinkler spacing, ft

The sprinkler spacing should allow for some overlap of sprinkler diameters. A spacing of about 80% of the wetted diameter should be adequate for OF. Using the design OF application rate and the above criteria for overlap, a sprinkler can be selected from a manufacturer's catalog.

Move-Stop Sprinkler Systems

With move-stop systems, sprinklers (or a single sprinkler) are operated at a fixed position in the field during application. After the desired amount of water has been applied, the system is turned off and the sprinklers (or sprinkler) are moved to another position in the field for the next application. Multiple sprinkler move-stop systems include portable hand-move systems, end tow systems, and side-wheel roll (also known as side-roll or wheel-line) systems. Single sprinkler move-stop systems include stationary gun systems.

Portable Hand-Moved Systems: Portable hand-moved systems consist of a network of surface aluminum lateral pipes connected to a main line which may be portable or permanent. The major advantages of these systems include low capital costs and adaptability to most field conditions and climates. They may also be removed from the fields to avoid interference with farm machinery. The principal disadvantage is the high labor requirement to operate the system.

End Tow Systems: End tow systems are multiple-sprinkler laterals mounted on skids or wheel assemblies to allow a tractor to pull the lateral intact from one position along the main to the next. The pipe and sprinkler design considerations are identical to those for portable pipe systems with the exception that pipe joints are stronger than hand moved systems to accommodate the pulling requirements.

The primary advantages of an end tow system are lower labor requirements than hand moved systems, relatively low system costs, and the capability to be readily removed from the field to allow farm emplements to operate. Disadvantages include crop restrictions to movement of laterals and cautious operation to avoid crop and equipment damage.

Side Wheel Roll: Side wheel roll or wheel move systems are basically lateral lines of sprinklers suspended on a series of wheels. The lateral line is aluminum pipe, typically 4 to 5 in. in diameter and up to 1,320 ft long. The wheels are aluminum and are 5 to 7 ft in diameter (see Figure 9-9). The end of the lateral is connected by flexible hose to hydrants located along the main line. The unit is stationary during application and is moved between applications by an integral engine powered drive unit located at the center of the lateral.

Figure 9-9 Side wheel roll sprinkler system.

The principal advantages of side wheel roll systems are relatively low labor requirements and overall costs, and freedom from interference with farm implements. Disadvantages include restrictions to crop height and field shape, and misalignment of the lateral caused by uneven terrain.

Stationary Gun Systems: Stationary gun systems are wheel-mounted or skid-mounted single sprinkler units, which are moved manually between hydrants located along the laterals. The advantages of a stationary gun are similar to those of portable pipe systems with respect to capital costs and versatility. In addition, the larger nozzle of the gun-type sprinkler is relatively free from clogging. The drawbacks to this system are similar to those for portable pipe systems in that labor requirements are high due to frequent sprinkler moves. Power requirements are relatively high due to high pressures at the nozzle, and windy conditions adversely affect distribution of the fine droplets created by the higher pressures.

Design Procedures: The design procedures regarding application rate, sprinkler selection, sprinkler and lateral spacing, and lateral design for move-stop systems are basically the same as those described for solid set sprinkler systems. An additional design variable for move-stop systems is the number of units required to cover a given area. The minimum required number of units is a function of the area convered by each unit, the application frequency, and the period of application. More than the minimum number of units can be provided to reduce the number of moves required to cover a given area. The decision to provide additional units must be based on the relative costs of equipment and labor.

Continuous Move Systems

Continuous move sprinkler systems are self-propelled and move continuously during the application period. The three types of continuous move systems are (1) traveling gun, (2) center pivot, and (3) linear move.

Traveling Gun Systems: Traveling gun systems are self-propelled, single large gun sprinkler units that are connected to the supply source by a hose 2.5 to 5 in. in diameter. Two types of travelers are available, the hose drag-type and the reel-type. The hose drag traveler is driven by a hydraulic or gas-driven winch located within the unit, or a gas-driven winch located at the end of the run. In both cases, a cable anchored at the end of the run guides the unit in a straight path during the application. The flexible rubber hose is dragged behind the unit. The reel-type traveler (see Figure 9-10) consists of a sprinkler gun cart attached to a take-up reel by a semi-rigid polyethylene hose. The gun is pulled toward the take-up reel as the hose is slowly wound around the hydraulic powered reel. Variable speed drives are used to control travel speeds. Typical lengths of run range between 660 and 1,320 ft, and spacings between travel lanes range between 165 and 330 ft. After application on a lane is complete, the unit shuts off automatically. Some units also shut off the water supply automatically. The unit must be moved by tractor to the beginning of the next lane.

Figure 9-10 Reel type traveling gun sprinkler.

The more important advantages of a traveling gun system are low labor requirements and relatively clog-free nozzles. They may also be adapted to fields of somewhat irregular shape and topography. Disadvantages are high power requirements, hose travel lanes required for hose drag units for most crops, and drifting of sprays in windy conditions.

In addition to the application rate and depth of application, the principal design parameters for traveling guns are the sprinkler capacity, spacing between travel lanes, and the travel speed.

The minimum application rate of most traveling gun sprinklers is about 0.23 in./h, which is higher than the infiltration rate of the less permeable soils. Therefore, the use of traveling guns on soils of low permeability without a mature cover crop is not recommended. The relationship between sprinkler capacity, lane spacing, travel speed, and depth of application is given by the following Equation:

$$D = \frac{q_s C}{(S_t)\ (S_p)} \qquad (9\text{-}10)$$

where D = depth of water applied, in.

q_s = sprinkler capacity, gpm

S_t = space between travel lanes, ft

S_p = travel speed, ft/min

C = conversion constant, 1.60

The typical design procedure is as follows:

1. Select a convenient application period, h/d, allowing at least 1 h between applications to move the gun.
2. Estimate the area to be irrigated by a single unit. This value should not exceed 80 acres.
3. Calculate the sprinkler discharge capacity using Equation 9-11.

$$q_s = \frac{(435)(D)(A)}{C\ t} \qquad (9\text{-}11)$$

where q_s = sprinker discharge capacity, gpm

D = depth of wastewater applied per application, in.

A = area irrigated per unit, acres

C = cycle time between applications, d

t = operating period, h/d

4. Select a sprinkler size and operating pressure from manufacturer's performance tables that will provide the estimated discharge capacity.

5. Calculate the application rate using Equation 9-12.

$$R = \frac{96.3\ Q}{\pi r^2} \qquad (9\text{-}12)$$

where R = application rate, in./h

Q = Sprinkler capacity, gpm

r = sprinkler wetted radius, ft

6. Compute the lane spacing as a percentage of the wetted diameter against spacing criteria in Table 9-10.

7. Adjust sprinkler selection and lane spacing as necessary to be compatible with soil intake rate.

Table 9.10: Recommended Maximum Lane Spacing for Traveling Gun Sprinklers

Wind Speed (mph)	Lane Spacing (% of wetted diameter)
0	80
0-5	70-75
5-10	60-65
>10	50-55

8. Calculate the travel speed using Equation 9-10 as rearranged:

$$S_p = \frac{1.6 q_s}{D\ S_t}$$

9. Calculate the area covered by a single unit.

$$A = \frac{S_t(\text{travel distance, ft/d})(\text{cycle, d})}{43{,}560\ \text{ft}^2/\text{acre}}$$

10. Determine the total number of units required.

$$\text{Units required} = \frac{\text{field area}}{\text{unit area}}$$

11. Determine the system capacity, Q

$$Q = (q_s)(\text{number of units})$$

<u>Example 9-3: Establish Preliminary Design Criteria for Reel Type Traveling Gun System</u>

<u>Conditions</u>: Loam soil, infiltration rate : 0.4 in/h, depth of wastewater applied, D : 3 in, field area : 100 acres, application cycle : every 10 d, average wind speed : 5 mph.

<u>Solution</u>:

1. Select a 15 h/d application period
2. Estimate 25 acres/unit
3. Calculate the sprinkler discharge capacity

$$q_s = \frac{(435)(3)(25)}{(10)(15)}$$

$$= 217.5\ \text{gpm}$$

4. Select a sprinkler with a 230 gpm capacity and a wetted diameter of 340 ft
5. Calculate the application rate

$$R = \frac{96.3(230)}{(170)^2}$$

$$= 0.24\ \text{in/h}\ (\quad 0.4\ \text{in/h, OK})$$

6. Lane spacing should be less than 70% to 75% of wetted diameter

$$S_t = 0.7\ (340)$$

$$= 238\ \text{ft}$$

use 240 ft

7. Calculate the travel speed

$$S_p = \frac{(1.6)(230)}{(3)(240)}$$

$$= 0.5 \text{ ft/min}$$

8. Calculate the area covered by a single unit

$$A = \frac{(240)(0.5)(15 \text{ h})(60 \frac{\text{min}}{\text{h}})(10 \text{ d})}{43{,}560}$$

$$= 24.8 \text{ acres}$$

9. Calculate the number of units required

$$\text{Units required} = \frac{100 \text{ acres}}{24.8 \text{ acres/unit}}$$

$$= 4.03$$

use 4 units

10. Calculate the system capacity, Q

$$Q = (q_s)(\text{number of units})$$

$$= (230 \text{ gpm})(4)$$

$$= 920 \text{ gpm}$$

Center Pivot Systems: Center pivot systems consist of a lateral with multiple sprinklers or spray nozzles that are mounted on self-propelled, continuously moving tower units (see Figure 9-11) rotating about a fixed pivot in the center of the field. Sprinklers on the lateral may be high pressure impact sprinklers; however, the trend is toward use of low pressure spray nozzles to reduce energy requirements. Water is supplied by a buried main to the pivot, where power is also furnished. The lateral is usually constructed of 6 to 8 in. steel pipe 200 to 2,600 ft in length. A typical system with a 1,288 ft lateral covers a 160 acre parcel. The circular pattern reduces coverage to about 130 acres, although systems with traveling end sprinklers or high pressure corner guns are available to irrigate the corners.

The tower units are driven electrically or hydraulically and may be spaced from 80 to 250 ft apart. The lateral is supported

10

Process Design—Slow Rate Systems

SYSTEM TYPES

Slow rate (SR) land treatment involves the controlled application of wastewater to a vegetated land surface. There are three basic types of SR systems:

1. Optimum hydraulic loading, ie: apply the maximum amount of water to the least possible land area; a "treatment" system.
2. Optimum irrigation potential, ie: apply the least amount of water to the maximum possible land area, or design for the production of a high market value crop; an "irrigation" system with treatment potential.
3. A toxic or hazardous component is present in the wastewater, requiring special measures for design, construction and operation.

Many of the system components (vegetation, preapplication treatment, transmission, distribution, etc.) may be identical for all three types. However, the land area utilized and the operational procedures will not be the same so it is necessary to develop a unique design approach for each case.

In general, industrial operations with easily degraded wastes the pivot circle. Areas of slower permeability can be flooded, causing crop damage and traction problems for the drive wheels. This particular problem has been encountered at the Muskegon project.

Wastewater application rates along a center pivot are determined by nozzle sizes and pressures, sprinkler spacing, length of lateral and type of sprinkler. The application rate is not affected by rotational speed of the center pivot. Rotational speed only affects the duration of application and the total depth of wastewater applied.

The flow capacity of a center pivot system is given by Equation 9-12.

$$Q = 1{,}890\ C\ A \tag{9-12}$$

where Q = flow capacity, gpm

C = wastewater application, in/d

A = area of application, acres

Since the water application rate pattern of a center pivot lateral is elliptical the maximum application rate is given by Equation 9-13 (11).

$$R = \frac{122.5\ Q}{r_1 r_2} \tag{9-13}$$

where R = maximum application rate of the last sprinklers, in/h

Q = center pivot capacity, gpm

r_1 = wetted radius of the center pivot lateral, ft

r_2 = wetted radius of the last few sprinklers, ft

A variety of sprinkler spacing packages are available from the manufacturers along with variable sizing of sprinkler nozzle sizes. The selection of the sprinkler package should take into account the soil infiltration rate, wind conditions, potential for soil compaction, and pressure requirements.

A limitation of center pivots is mobility under certain soil conditions. Some clay soils can build up on wheels and eventually cause the unit to stop. Drive wheels can lose traction on slick (silty) soils and can sink into soft soils and become stuck. As a result, high flotation tires are used and low tire pressures are recommended according to the data in Table 9-11.

Table 9.11: Recommended Soil Contact Pressure for Center Pivots

Percent Fines	Pounds per Square Inch
20	25
40	16
50	12

Note: To illustrate the use of this table, if 20% of the soil fines pass through a 200 mesh screen, the contact pressure of the supporting structure to the ground should be no more than 25 lb/in^2. If this is exceeded, one can expect wheel tracking problems to occur.

Linear Move Systems: Linear move systems are constructed and driven in a similar manner to center pivot systems, except that the unit moves continuously in a linear path rather than a circular path. Complete coverage of rectangular fields can thus be achieved while retaining all the advantages of a continuous move system. Water can be supplied to the unit through a flexible hose that is pulled along with the unit or it can be pumped from an open center ditch constructed down the length of the linear path. Slopes greater than 5% restrict the use of center ditches. Manufacturers should be consulted for design details.

REFERENCES

1. Design of Wastewater and Stormwater Pumping Stations, Water Pollution Control Federation, Manual of Practice No. FD-4 (1981).
2. Hydraulic Institute Standards, Hydraulic Institute, Cleveland, Ohio (1983).
3. Hart, W.E., Irrigation System Design, Colorado State University, Department of Agricultural Engineering, Ft. Collins, Colorado (November 10, 1975).

4. Booher, L.J., Surface Irrigation, FAO Agricultural Development Paper No. 94, Food and Agricultural Organization of the United Nations, Rome (1974).

5. Merriam, J.L. and Keller, J., Irrigation System Evaluation: A Guide for Management, Utah State University, Logan, Utah (1978).

6. McCulloch, A.W. et al, Lockwood-Ames Irrigation Handbook, Lockwood Corporation, Gering, Nebraska (1973).

7. Border Irrigation, Irrigation, Chapter 4, SCS National Engineering Handbook, Section 15, U.S. Department of Agriculture, Soil Conservation Service, (August 1974).

8. Fry, A.W., and Gray, A.S., Sprinkler Irrigation Handbook, Rain Bird Sprinkler Manufacturing Corporation, Glendora, California, 10th Edition (1971).

9. Sprinkler Irrigation, Irrigation, Chapter 11, SCS National Engineering Handbook, Section 15, U.S. Department of Agriculture, Soil Conservation Service (July 1968).

10. Pair, C.H. et al, Irrigation, Fifth Ed., Silver Spring, Maryland, Irrigation Association (1983).

11. Dillon, R.C., Hiler, E.A., and Vittetoe, G., Center-pivot Sprinkler Design Based on Intake Characteristics, ASAE Trans. 15: 996-1001 (1972).

10

Process Design—Slow Rate Systems

SYSTEM TYPES

Slow rate (SR) land treatment involves the controlled application of wastewater to a vegetated land surface. There are three basic types of SR systems:

1. Optimum hydraulic loading, ie: apply the maximum amount of water to the least possible land area; a "treatment" system.
2. Optimum irrigation potential, ie: apply the least amount of water to the maximum possible land area, or design for the production of a high market value crop; an "irrigation" system with treatment potential.
3. A toxic or hazardous component is present in the wastewater, requiring special measures for design, construction and operation.

Many of the system components (vegetation, preapplication treatment, transmission, distribution, etc.) may be identical for all three types. However, the land area utilized and the operational procedures will not be the same so it is necessary to develop a unique design approach for each case.

In general, industrial operations with easily degraded wastes and municipalities in the humid parts of the country will seek to minimize land and distribution system costs, and will implement the first type of SR system. In the arid parts of the world where the water itself has a significant economic value it is often cost effective to design a type 2 system. Whenever toxic or hazardous materials are present, the type 3 system is necessary regardless of other conditions or opportunities.

The design of type 1 and 3 systems is always based on the Limiting Design Parameter (LDP) concept defined in Chapters 2 and 3. The LDP for typical municipal effluents and many industrial wastewaters will either be the hydraulic capacity of the soil in limiting water movement or nitrogen leakage to drinking water aquifers. The LDP for some industrial wastes may be metals, solids, organics or one of the other materials discussed in detail in Chapter 3.

The design of type 2 irrigation systems is based on the water needs of the crop to be grown and is similar to standard irrigation system design. However, it is necessary to check to insure that an LDP is not being exceeded. The LDP in this case will usually apply to any effects on the quantity or quality of the crop to be grown, or to nitrogen impacts on the groundwater. In general, the application rates for type 2 irrigation are usually much lower than the ability of the soil to transmit water, so hydraulic capacity of the soil is not typically a constraint.

MAXIMUM HYDRAULIC LOADING RATE, L_w

In all cases the maximum hydraulic loading rate, as controlled by soil permeability should be determined to establish the capacity of the soil profile to transmit water and to determine if this factor is the LDP for design. The hydraulic loading rate is the volume of wastewater applied per unit area of land over at least one loading cycle; and is commonly expressed in inches/week, inches/month or feet/year.

The general site water balance, with runoff of applied wastewater assumed to be zero, is given by:

$$L_w = ET - P_r + P_w \tag{10-1}$$

where L_w = wastewater hydraulic loading rate

ET = evapotranspiration rate

P_r = precipitation rate (water equivalent in snow fall periods)

P_w = design percolation rate

Common units (inch, foot, etc.) are required for all factors in the equation. A common time base for all factors is also necessary. An annual basis is suitable for preliminary screening but a monthly basis should be used for final design. In both cases the L_w is used to compute the area required for SR treatment.

Example 10-1: Determine Minimum Land Area Based on Hydraulic Loading Criteria.

Conditions: An industry discharges 20,000 gpd. For preliminary screening find the minimum number of acres required to accept the wastewater if: ET = 3 ft/yr, P_r = 5 ft/yr, P_w = 6 ft/yr.

Solution:

1. $L_w = ET - P_r + P_w$

$= (3) - (5) + (6)$

$= 4$ ft/yr

2. $Q = (20{,}000 \text{ gal/d})(0.1337 \text{ ft}^3/\text{gal})(365 \text{ d/yr})$

$= 976{,}010 \text{ ft}^3/\text{yr}$

3. $\text{Minimum Area} = \frac{Q}{L_w} = \frac{976{,}010}{(4)\ (43{,}560 \text{ ft}^2)/\text{acre}} = 5.6 \text{ acres}$

Design Percolation Rate, P_w

The design percolation rate, (P_w) in Equation 10-1 is a function of the type of system being designed. If it is a type 1 system designed for optimum hydraulic loading then P_w is a function of the limiting permeability in the soil profile. If it is an irrigation system (type 2) then P_w is the amount of water required to leach salts out of the root zone so plant growth will not be inhibited. Both of these approaches are described in detail below. Type 3 systems may sometimes be designed for $P_w = 0$.

Type 1 Systems - Permeability Limiting: The design P_w is taken as a conservative percentage of the limiting hydraulic capacity determined by field tests using the procedures described in Chapter 7. The top 5 to 8 ft of the soil profile is the depth of concern. If the minimum soil permeabilities are variable over the site, determine a weighted average based on soil type.

Figure 4-6 can then be used to estimate the P_w on an annual basis. The values on the figure range from 4 to 10% of the clean water permeability of the soil. The P_w on a daily basis can also be calculated with:

$$P_w\ (\text{daily}) = (K,\ \text{in/h})(24\ \text{h/d})(0.04\ \text{to}\ 0.10) \qquad (10\text{-}2)$$

where K = limiting saturated hydraulic conductivity or permeability of site, in/h

The percentage to use in the calculation is a judgement decision by the designer. If the soils are relatively uniform and sandy ($K \geq 2$ in/h) the upper 10% limit would be appropriate. The lower limit should be used for finer textured soils ($K \leq 0.6$ in/h) and when there is considerable variability in soil type on the site.

The monthly P_w is then:

$$P_{w\ (\text{month})} = (P_{w\ (\text{daily})})(\text{operating day per month}) \qquad (10\text{-}3)$$

The number of operating days in a particular month may depend on:

- Crop management. Downtime must be allowed for harvesting, planting, and cultivation as applicable.

- Precipitation. Downtime for precipitation is already factored into the water balance computation. No further adjustments are necessary.
- Freezing temperatures. Subfreezing temperatures may cause soil frost that reduces surface infiltration rate. Operation is usually stopped when this occurs. The most conservative approach to adjusting the monthly percolation rate for freezing conditions is to allow no operation for days during the month when the mean temperature is less than 32° F. A less conservative, but acceptable, approach is to use a lower minimum temperature. The recommended lowest mean temperature for operation is 25° F. Data sources and procedures for determining the number of subfreezing days during a month are discussed in Chapters 6 and 8. Nonoperating days due to freezing conditions may also be estimated using the EPA-1 computer program without precipitation constraints. For forested sites, operation can often continue during subfreezing conditions.
- Seasonal crops. When a single annual crop is grown, wastewater is not normally applied during the winter season, although applications may occur after harvest and before the next planting.

Chapter 8 demonstrates the procedures to be used to determine storage days, based on climatic factors. The additional agronomic factors listed above can be determined from local experience in the area once the type of cover vegetation is tentatively identified. It is necessary to select the vegetation type at an early stage of design so that the crop uptake of nitrogen or other constituents can be estimated.

Type 2 Systems - Optimize Irrigation Potential: The design wastewater percolation rate P_w in this case is usually zero in the wet months when the natural precipitation exceeds ET. In the dry months P_w is equal to the leaching requirement (LR) which is that volume of water needed to move salts out of the root zone. The leaching requirement on an annual basis (expressed as a percentage) is determined by:

$$LR = \frac{P_w}{L_w + (\text{Excess } P_r)} (100) \qquad (10\text{-}4)$$

where LR = leaching requirement, %

other terms defined previously.

The LR is dependent on the salt tolerance of the least salt tolerant crop to be grown (1) as detailed in Chapter 3 (Equation 3-5).

Figure 10-1 can be used to determine the LR for a variety of crops depending on the salinity of the irrigation water. It ranges from 2% for non sensitive crops and low salinity waters to over 30% for high salinity waters and sensitive crops. See Chapter 3 for a more detailed discussion of salinity factors.

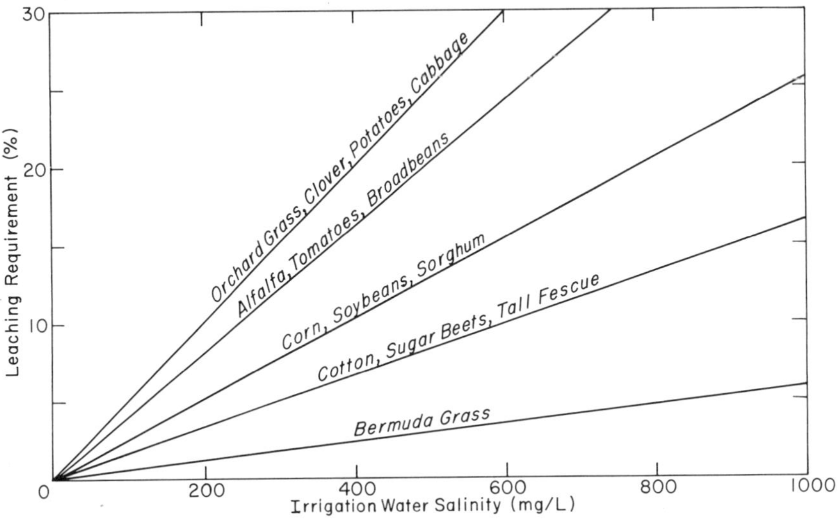

Figure 10-1 Leaching requirement vs salinity for various crops, After (1).

In arid climates there is typically no excess P_r available for deep percolation in the dry months so:

$$P_w = (\frac{LR}{100})(ET - P_r) \tag{10-5}$$

Substitution in the original water balance Equation (10-1) yields:

For periods when $(ET - P_r) \leqslant 0$

$$L_w = (ET - P_r)(1 + \frac{LR}{100}) \tag{10-6}$$

A further modification is necessary to account for water losses to percolation and evaporation in the conveyance and distribution systems. This overall efficiency ranges from about 65% to over 85% (2). The final water balance equation for the irrigation case (type 2 system) is:

$$L_w = (ET - P_r)(1 + \frac{LR}{100})(\frac{100}{E_s}) \tag{10-7}$$

where E_s = efficiency of distribution system, %
(65% - 75% for surface systems, 70% - 80% for sprinklers)
other terms defined previously.

Design Precipitation Rate, P_r

An estimate on an annual basis is suitable for preliminary calculations during site planning. Monthly values are needed for final design. These should be based on a 5 year return period frequency analysis for monthly precipitation. An alternate approach uses a 10 year return period for annual precipitation. These values are then distributed monthly based on the ratio of average monthly to average annual precipitation.

Design Evapotranspiration Rate ET

This factor defines the consumptive use of water by the surface vegetation. It is a critical component in the water balance for both crop production and water quality concerns. In the latter case, a high water loss due to ET will tend to increase the concentration of substances in the remaining percolate. The potential evapotranspiration ET_p is defined as the water loss that could occur from a vegetated field (typically grass covered) with soil water readily available to the plants and with the plants in a vigorous growth stage. See Chapter 5 for discussion and procedures for estimating ET for a particular crop.

A preliminary estimate of ET can also be obtained with Holdridge's (3) method:

$$ET_p = 1.07\ T_m - 34.24 \qquad (10\text{-}8)$$

where ET_p = potential evapotranspiration, in./mo

T_m = mean monthly air temperature, ^oF

In humid regions these estimates of ET_p are usually sufficient for design when perennial, full cover crops are to be used.

HYDRAULIC LOADING RATE BASED ON THE LDP

In many cases the constituent LDP for municipal effluents will be nitrogen to protect drinking water aquifers at the project boundary. Industrial wastes may have one of the other materials discussed in detail in Chapter 3 as the LDP. The calculation procedure is derived below in terms of nitrogen but the same approach is valid for other elements or materials by substitution of the appropriate boundary conditions. The mass balance for wastewater nitrogen on the site is given by:

$$L_n = U + D + 2.7\ C_p P_w \qquad (10\text{-}9)$$

where L_n = mass loading of nitrogen, lb/acre·yr

U = crop uptake, lb/acre·yr

D = site losses, volatilization, denitrification, lb/acre·yr

C_p = percolate nitrogen concentration, mg/L

P_w = percolation quantity, ft/yr

The site losses D are a function of the amount of nitrogen applied:

$$D = f\ (L_N)$$

Substituting for D in Equation (10-9):

$$L_n = U + f\ (L_n) + 2.7\ C_pP_w$$

Solving for P_w:

$$P_w = \frac{(1-f)L_n - U}{2.7\ C_p} \qquad (10\text{-}10)$$

The water balance on the site is given by:

$$L_{wn} = ET - P_r + P_w \qquad (10\text{-}11)$$

where L_{wn} = hydraulic loading controlled by nitrogen as the LDP, ft/yr

The amount of nitrogen in the annual hydraulic loading L_{wn} is:

$$L_n = 2.7\ C_nL_{wn} \qquad (10\text{-}12)$$

where C_n = concentration of nitrogen in the applied wastewater. (See Chapter 8 for determining C_n after lagoon treatment or storage ponds.)

Rearranging Equation (10-11) and solving for P_w:

$$P_w = L_{wn} + (P_r - ET) \qquad (10\text{-}13)$$

Setting Equation (10-10) and (10-13) equal to each other:

$$L_{wn} + (P_r - ET) = \frac{(1-f)(L_n) - U}{2.7C_p}$$

Then substitute Equation (10-12) on the right side for L_n:

So:

$$L_{wn} + (P_r - ET) = \frac{(1-f)(2.7C_n)(L_{wn}) - U}{2.7C_p}$$

Finally, combine terms and solve for L_{wn}:

$$L_{wn} = \frac{C_p(P_r - ET) + 0.37U}{(1-f)\ C_n - C_p} \qquad (10\text{-}14)$$

(Note: The coefficient 0.37 is based on the use of feet as

the dimensional unit for $(P_r\text{-}ET)$. If inches are used the coefficient becomes 4.4; if centimeters are used then the coefficient is 10.)

Equation (10-14) can be used to determine the hydraulic loading allowed for a particular wastewater and a specified combination of site factors (P_r ET, U) and regulatory requirements (C_p). The regulatory constraint when nitrogen is the LDP is the nitrate concentration in the groundwater at the project boundary. To insure a conservative design the C_n and C_p values in the equation are taken as the total nitrogen present, not just the nitrate fraction since it is possible that other forms of N may be oxidized to nitrate in the soil profile. The equation is also very conservative since it is based on the concentration C_p in the percolate prior to any mixing or dispersion in the groundwater. It will be advantageous for large projects, particularly in arid climates to determine the degree of mixing, dispersion and dilution that will occur between the application point and the project boundary. In that case C_p' would be equal to 10 mg/L nitrogen at the project boundary. An allowable C_p in the wastewater percolate can then be determined and Equation (10-14) solved for the allowable hydraulic loading.

The f factor is discussed in Chapter 3 and ranges from zero to 50% depending on waste characteristics and application methods. For food processing wastewaters with high (>5) BOD : N ratios, f values of up to 0.8 can be realized. The f for primary and lagoon effluents will be about 0.2; for secondary treatment it should range from 0.1 to 0.2. Highly oxidized, filtered effluents would have an f of 0.1 or less. See Chapter 16 for similar criteria for sludges.

Design Modification for Supplemental Nitrogen

In some cases, supplemental nitrogen, in addition to that contained in the wastewater is also applied to the site. This could be in the form of commercial fertilizers, or manures from grazing animals or periodic sludge applications from the preapplication treatments. The Nitrogen section in Chapter 3 discusses these sources in detail, and presents the calculation methods to be used in determining the N contribution from manures, sludges and other solid wastes. If supplemental nitrogen from any source is to be added then Equation (10-14) must be modified accordingly:

$$L_{wn} = \frac{C_p(P_r - ET) + 0.37(U - S_n)}{(1-f)(C_n) - C_p} \qquad (10\text{-}15)$$

where S_n = supplemental nitrogen, lb/acre·yr

other terms as defined previously and $(P_r\text{-ET})$ in ft.

EXAMPLE 10-2: Determine Allowable Hydraulic Loading if Nitrogen is the LDP.

Conditions:

U = 500 lb/acre·yr (coastal bermudagrass, from Chapter 5)

$(P_r\text{-ET})$ = -1.0ft/yr (a dry climate)

C_n = 50 mg/L total N (a strong wastewater)

f = 0.20 (assume lagoon effluent)

C_p = 10 mg/L (required by State)

S_n = 0

Solution:

1. $$L_{wn} = \frac{C_p\,(P_r\text{-ET}) + 0.37\,U}{(1-f)(C_n) - C_p}$$

$$= \frac{(10)(-1) + (0.37)(500)}{(0.8)(50) - 10}$$

= 5.8 ft of water per year.

2. The calculation for final design would be repeated on a monthly basis to insure that sufficient water

is applied in the dry months and the C_p requirement is satisfied at all times.

3. Maintaining a C_p of 10 mg/L in arid climates is difficult because of the concentrating effects of the high ET losses. Repeating this example for a more arid location, but all other factors the same, demonstrates the concern:

$$\text{Assume:} \quad (P_r\text{-ET}) = -5 \text{ ft/yr}$$

$$\text{Then:} \quad L_{wn} = \frac{(10)(-5) + (0.37)(500)}{(0.8)(50) \quad -10}$$

$$= 4.5 \text{ ft/yr}$$

4. This is the maximum amount of wastewater that could be applied and still maintain a C_p of 10 mg/L. However, the crop in this example needs at least 5 ft of water per year to survive. Supplemental irrigation water will be required to make up the deficit.

5. Equations (10-14) and (10-15) are only valid for positive values of P_w. However, P_w should be positive in all cases. In arid climates the leaching requirement LR may control design and it may be necessary to use low nitrogen water sources for this purpose if the C_p is a regulatory requirement.

6. In step 3 above the water deficit was 5 ft/yr. Assuming the LR was 10% the hydraulic loading for an irrigation operation would be:

$$L_w = (5)(1.10) = 5.5 \text{ ft/yr}$$

To maintain the specified C_p at 10 mg/L only 4.5 ft of this could be wastewater, the remainder would have to come from other sources.

LDP for Materials Other than Nitrogen

The basic approach described above for nitrogen is valid for any other substance also. For example, assume that a small industry involved with galvanized metal products is interested in land treatment for its zinc laden effluent. The mass balance for this case is given by a modification of Equation 10-9:

$$L_{zn} = U + D + SA + 2.7\, C_p P_w$$

where L_{zn} = mass loading of zinc, lb/acre·yr

U = probable crop uptake, lb/acre·yr (see ref. 6 for typical values)

D = site losses, assume = 0 for non volatile substances

SA = soil profile accumulation, lb/acre·yr. Example: assume a soil CEC 15, then recommended lifetime limit for Zn = 1,000 lb/acre (See Table 3-10).

C_p = allowable concentration in percolate, mg/L U.S. EPA standards = 5.0 mg/L for zinc in drinking water

P_w = volume of percolate, ft/yr

Use Equation 10-12, with zinc specified as the parameter of concern:

$$L_{zn} = 2.7\, C_{zn}\, L_{w_{zn}}$$

Then combine Equation 1-12 and 10-16 and rearrange terms as before to determine the L_w limited by zinc:

$$L_{w_{zn}} = \frac{C_p(P_r - ET) + 0.37U + 0.37SA}{C_{zn} - C_p}$$

In theory, this is the allowable annual waste loading to maintain the specified 5 mg/L, C_p in the percolate. In fact, since zinc and other metals are strongly adsorbed by the soil profile there may be none at all in the percolate until the assimilative capacity of the profile is reached. In such cases the limiting hydraulic loading should be based on only the crop removal and the soil profile accumulation:

$$L_{w_{zn}} = \frac{(U + SA)}{2.7\ C_{zn}} \qquad (10\text{-}16)$$

All terms defined previously.

Chapter 3 should be consulted to determine the anticipated responses for the substance of concern to insure that all factors are included in the mass balance equation. In general, the form of the hydraulic loading equation will be similar to either Equation (10-15) or (10-16). Wastewaters or sludges with a toxic or hazardous component typically use an equation similar to (10-16) for design.

MONTHLY WATER BALANCE AND HYDRAULIC LOADING RATE FOR FINAL DESIGN, L_{wD}

The allowable hydraulic loading based on the LDP should be compared to the maximum possible hydraulic loading (L_w) based on soil permeability. The lowest of these two values then controls design. A monthly water balance is then prepared to determine the specific monthly hydraulic loadings for design. Table 10-1 illustrates a typical water balance for a site in an arid climate, using Equations (10-1) and (10-2) for a soil with a moderately slow permeability (K = 0.2-0.6 in/h) so:

$$\begin{aligned} \text{Maximum } P_w &= (K)\ (24\ h/d)(.05)(30\ d/mo) \\ &= (0.2)(24)(0.05)(30) \\ &= 7.2\ in/mo \end{aligned}$$

Operating time, assume:

April - Oct.	=	30d each	so:	P_w = 7.2 in/mo
Nov.	=	28d	so:	P_w = 6.2 in/mo
Dec.	=	23d	so:	P_w = 5.6 in/mo
Jan.-Feb.	=	10d each	so:	P_w = 2.4 in/mo
March	=	27d	so:	P_w = 6.5 in/mo

Table 10.1: Typical Water Balance to Determine Maximum Hydraulic Loading L_w

Month	ET	P_r	$(ET\text{-}P_r)$	P_w	L_w
	(inches)				
Jan	0.9	1.0	-0.1	2.4	2.3
Feb	2.0	1.1	0.9	2.4	3.3
Mar	3.8	1.1	2.7	6.5	9.2
Apr	5.2	0.8	4.4	7.2	11.6
May	7.0	0.2	6.8	7.2	14.0
June	8.6	0.1	8.5	7.2	15.7
July	9.4	0.0	9.4	7.2	16.6
Aug	8.7	0.0	8.7	7.2	15.9
Sept	5.8	0.1	5.7	7.2	12.9
Oct	4.3	0.3	4.0	7.2	11.2
Nov	2.0	0.5	1.5	6.2	7.7
Dec	1.0	1.0	0.0	5.6	5.6
Annual	58.7	6.2	52.5	73.5	126

The values in Table 10-1 would be for a type 1 system where the intent is to maximize the hydraulic loading. A type 2 irrigation system would be designed to make up the water deficit $(ET\text{-}P_r)$ plus a leaching requirement. Assuming a 10% LR for this case would give:

$$L_w = (ET\text{-}P_r)(1 + \frac{LR}{100})$$

$$= (52.5)(1.1)$$

$$= 57.8 \text{ in/yr}$$

The monthly L_w values would then be calculated in the same way.

A water balance for a humid climate would have more negative values in the $(ET\text{-}P_r)$ column reflecting a net excess of precipitation. The L_w column would be the summation P_w and $(ET\text{-}P_r)$ as shown on Table 10-1. However, if a negative value results, a zero is entered in the L_w column since the wastewater hydraulic loading cannot be less than zero.

If the L_w based on soil permeability is less than the hydraulic loading allowed by nitrogen or some other constituent this water balance then becomes the basis for design. If nitrogen or some other substance controls then additional calculations are necessary.

Example 10-3: Establish the Design Hydraulic Loading L_{wD}

Conditions: Type 1 system, assume wastewater nitrogen, C_n = 25 mg/L, Crop uptake (U) = 300 lb/acre.

$$C_p = 10 \text{ mg/L}$$

$$f = 0.20$$

Arid Climate, annual $(P_r\text{-ET}) = -52.5$ in. (Table 10-1)

Humid Climate, assume annual $(P_r\text{-ET}) = +19.7$ in.

From Table 10-1, $P_w = 73.5$ in/yr

Find the design hydraulic loading for both sites.

Solution:

1. Arid climate conditions:

$$L_w = (ET\text{-}P_r) + P_w$$

$$= 52.5 + 73.5$$

$$= 126 \text{ in/yr}$$

$$= 10.5 \text{ ft/yr}$$

$$L_{wn} = \frac{C_p(P_r\text{-ET}) + 0.37U}{(1-f)(C_n) - C_p}$$

$$= \frac{(10)(-52.5/12) + 0.37\ (300)}{(0.8)(25) - 10}$$

$$= 6.7 \text{ ft/yr}$$

L_{wn} controls design, so $L_{wD} = 6.7$ ft/yr

2. Humid climate conditions:

$$L_w = 19.7 + 73.5$$

$$= 93.2 \text{ in/yr}$$

$$= 7.8 \text{ ft/yr}$$

$$L_{wn} = \frac{(10)(+\frac{19.7}{12}) + (0.37)(300)}{(0.8)(25) - 10}$$

$$= 12.8 \text{ ft/yr}$$

In this case the L_w based on soil permeability controls, so $L_{wD} = 7.8$ ft/yr.

If crop uptake (U) or supplemental nitrogen (S_n) are factors in the equation it is necessary to determine a monthly increment for each component. The application schedule for S_n is determined by the design, or if commercial fertilizers are involved, by local agronomic practice. In some cases as shown on Figure 5-3 a monthly value for U can be determined. In other cases when only the annual crop uptake is known the monthly value can be estimated by distributing the annual uptake to each month in the same proportion as that month's ET to the total ET for the growing season.

If the L_w based on soil permeability controls design then the monthly and annual values have already been determined by the preliminary water balance as shown on Table 10-1. If L_{wn} (or some other LDP) controls design it is necessary to use Equation (10-15) modified for the particular LDP. These monthly values are then compared to the previously calculated L_w values and the lower of the two used as the design hydraulic loading for a particular month.

Nitrogen loading is more likely to govern the design hydraulic loading rate for systems in arid climates than in humid climates. The reason for this is that the net positive ET rate in arid climates causes an increase in the concentration of the nitrogen level in the percolating water.

For systems in arid climates, it is possible that the design monthly hydraulic loading rates based on nitrogen limits will be less than the irrigation requirements of the crop. The designer should compare the design L_w with the irrigation requirement to determine if this situation exists. If it does exist, the designer has three options available:

1. Reduce the concentration of applied nitrogen (C_n) through preapplication treatment.
2. Demonstrate that sufficient mixing and dilution will occur with the existing groundwater to permit

higher values of percolate nitrogen concentration (C_p) to be used in Equation (10-15).

3. Select a different crop with a higher nitrogen uptake (U).

LAND AREA DETERMINATION

The land area required for wastewater treatment is based on the design hydraulic loading L_{wD}. This surface area, actually receiving wastewater, is called the field area and is determined by:

$$A_F = \frac{(C)(Q) + V_s}{L_{wD}} \tag{10-17}$$

where A_F = field area, acres

C = conversion factor = 3.06 acre·ft/mg

Q = average annual wastewater flow, mg/yr

= (average daily flow from community or industry) (365 d/yr)

L_{wD} = design annual hydraulic loading, ft/yr

V_s = net loss or gain of water in storage pond due to excess precipitation or evaporation and seepage, acre-ft/yr (see Equation 8-1).

An iterative approach to the calculations is necessary for SR systems that incorporate storage since there is an interrelationship between the storage area required and the hydraulic loading on the field area. The first calculation is made without considering the V_s factor to determine an approximate land area. The procedure is defined below:

1. Determine preliminary $A_F = \frac{(C)(Q)}{L_{wD}}$

2. Use monthly L_{wD} values and A_F to determine monthly volumetric applications:

$$W = (L_{wD})(A_F)$$

W = monthly storage pond withdrawal, acre·ft (see Chapter 8, Table 8-6).

3. Assume a storage pond depth and tabulate a monthly water balance for storage pond (see Table 8-6).
4. Determine net precipitation or net evaporation/seepage for assumed pond. This is V_S in Equation (10-17). If net precip. use (+) sign; if net loss use (-) sign.
5. Solve Equation (10-17) for A_F, including the V_S factor.
6. Repeat steps 2 to 4 above to insure a balance. Adjust assumed area or depth of storage pond as necessary.

Example 10-4: Field Area Determination

Conditions: Determine the preliminary field area for a 100,000 gpd industrial discharge if the annual L_{wD} is 6.7 ft/yr.

Solution:

$$Q \simeq \frac{(100{,}000 \text{ gpd})(365 \text{ d/yr})}{(1{,}000{,}000 \text{ g/mg})}$$

$$= 36.5 \text{ mg/yr}$$

$$\text{Preliminary } A_F = \frac{(C)(Q)}{L_{wD}}$$

$$= \frac{(3.06)(36.5)}{6.7}$$

$$= 16.7 \text{ acres}$$

The land area determined with Equation (10-17) is the treatment surface required. It does not include roads, buffer zones, maintenance buildings, or unusable portions of the site. An allowance of about 20% is made for these factors in preliminary design. The final design must include an exact determination for each of these requirements. Usable portions of a particular land parcel are delineated, their areas determined and summed up until the total equals the required A_F.

Buffer Zone Requirements

The objectives of buffer zones around land treatment sites are to control public access, and in some cases, improve project aesthetics. There are no universally accepted criteria for determining the width of buffer zones around SR treatment systems. In practice, the widths of buffer zones range from zero for remote systems to 200 ft or more for systems using sprinklers near populated areas. In many states, the width of buffer zones is prescribed by regulatory agencies and the designer should determine if such requirements exist.

The requirements for buffer zones in forest systems are generally less than those of other vegetation systems because forests reduce wind speeds and, therefore, the potential movement of aerosols. Forests also provide a visual screen for the public. A minimum buffer zone width of 50 ft should be sufficient to meet all objectives, if the zone contains trees with a dense leaf canopy.

Storage Requirements

See Chapter 8 for a detailed discussion and calculation procedures. When storage is a component in an SR system it may be advantageous not to bypass the pond in the application season to allow reductions in coliforms and nitrogen to occur as described in Chapter 8. Algal production in storage ponds should not affect SR operations.

Crop Selection

The type of crop selected will directly influence the land area required, if crop uptake is a factor in determining the design hydraulic loading L_{wD}. In most cases crop selection will be one of the first design decisions in SR design. See Chapter 5 for discussion and selection procedures.

Distribution System

It is necessary for type 2 irrigation systems to decide on the type of distribution system that will be used, at an early stage of design. The system efficiency (see Equation 10-7) is a significant factor in determining the L_w and the amount of land that can be irrigated. An early decision on distribution method is less critical for type 1 treatment systems.

APPLICATION SCHEDULING

A regular, routine application schedule is usually adopted for type 1 treatment systems for operational convenience. Sprinklers with an application rate of 0.2 to 0.3 in/h are often employed in SR systems. This will not usually exceed the intake rate of most soils, so runoff is prevented. It is then typical to operate the sprinkler unit continuously for a sufficient number of hours to achieve the design weekly loading. The application is then repeated seven days later. Operation can either be manual, automated with time switches or some combination.

The scheduling of a type 2 irrigation system is dependent on the climate and the crop to be grown. The purpose is to maintain sufficient moisture in the root zone to sustain plant growth. The water available for plant use is defined as the difference between the field capacity and the wilting point (see Chapters 4 and 5).

The usual range of the deficit that is allowed ranges from 30% to 50% of the available water in the root zone, depending on the type of crop and stage of growth (see Chapter 4, and Table 4-2 for further discussion of field capacity, etc.). An irrigation is scheduled when the soil moisture reaches the predetermined deficit. This can be estimated manually (see Table 4-2) or measured with tensiometers or similar devices. The latter can be used in a completely automated system to start-up, shut-down and shift applications from field to field. The amount of water to be applied in each irrigation can be determined with:

$$I_T = I_D \left[1 + \frac{LR}{100}\right]\left(\frac{100}{E_a}\right) \qquad (10\text{-}18)$$

where I_T = total depth of water to be applied during an irrigation, in.

I_D = soil moisture deficit to be replaced, in.

LR = leaching requirement, %

E_a = application efficiency, % (See Equation 10-7 for typical values)

Example 10-5: Determine Design Hydraulic Loading and Field Area for an Industrial Operation in the North Central U.S.

Conditions: 100,000 gpd design flow. BOD = 900 mg/L, TSS = 450 mg/L, Total N = 60 mg/L, Total P = 20 mg/L, K = 8 mg/L, Grease content 0.2%, Fecal Coliform = 10^6/100 ml.

250 acres are available adjacent to plant with silt loam soil (k = 0.5 in/h), and groundwater at 30 ft. EPA climate programs indicate 130 days of storage for site. Residential area next to treatment site. (P_r-ET) = + 2 in/yr for area. Type 1 treatment system proposed.

Solution:

1. Preliminary design assumes grass covered fields (U = 280 lb/acre·yr) and application with sprinklers.
2. Since 130 day storage is required, use a three cell lagoon for preliminary treatment/storage. Design in accordance with Chapter 8:

First Cell: Aerated, 10 ft deep, 12 h detention

$$\text{Surface Area} = \frac{(100{,}000)(12)}{(24)(7.48)(10)} = 668 \text{ ft}^2$$

Net precipitation on cell:

$$V_s = \left(\frac{2.0}{12}\right)(668) = 111.3\ ft^3/yr$$

Second and Third Cells: Non aerated, variable depth (12 ft max) 130 d storage capacity.

$$\text{Total Surface} = \frac{(100,000)(130)}{(7.48)(12)} = 144,831\ ft^2 = 3.32\ \text{acres}$$

area of each cell = 1.66 acres

2 in. net precipitation, $V_s = \left(\frac{2}{12}\right)(144,831) = 24,139\ ft^3/yr$

Total V_s, all cells = + 24,250 ft^3/yr

Storage period = December 1 - April 10

Discharge period = April 11 - November 30

= 365-130 = 235 d/yr

= 33.6 wk/yr

$$\text{Average weekly discharge} = \frac{(100,000)(365)/(7.48) + 24,250}{33.6}$$

$$= 145,950\ ft^3/wk$$

3. Determine nitrogen removal in pond system.

a. During filling and storage period: water temperature = 8^oC, pH = 7.5, "average" detention time = $\frac{130}{2} = 65$ d, $k_T = (0.0064)(1.039)^{(8-20)} = 0.004$

Equation (8-5) Pond Effluent N at end of storage period

$$= (60)e^{-0.004\left[65 + 60.6(7.5-6.6)\right]}$$

$$= 37.2\ mg/L$$

b. During discharge period: assuming plug flow conditions it will take about 83 d to empty the stored water. Temperature = 15° C, pH = 8.5, $k_T = (0.0064)(1.039)^{(15-20)} = 0.0053$.

So: Average pond effluent during first 83 d of discharge

$$= (37.2)e^{-0.0053\left[83 + 60.6(8.5-6.6)\right]}$$

$= 13.0$ mg/L total N

Average pond effluent during final 152 d of discharge. Temperature = $10^\circ C$, pH = 8, $k_T = (0.0064)(1.039)^{(10-20)} = 0.0044$

$$\text{Effluent N} = (60)e^{-0.0044\left[130 + 60.6(8-6.6)\right]}$$

$$= 23.3 \text{ mg/L total N}$$

$$\text{Average effluent N during discharge period} = \frac{(13)(83) + 23.3(152)}{235}$$

$$= 19.7 \text{ mg/L total N}$$

4. Determine fecal coliform removal in ponds, using Equation (8-7).

   a. During filling and storage period, t = 65 d, T = $8^\circ C$, use $k_{FC} = 0.03$.

$$FC_f = \frac{FC_i}{e^{(t)(k_{FC})(\theta)^{(T-20)}}}$$

$$= \frac{10{,}000{,}000}{e^{(65)(0.013)}}$$

$$= 4{,}288{,}548 = 4 \times 10^6/100 \text{ ml}$$

   b. During initial discharge period, t = 83d, T = $15^\circ C$, $k_{FC} = 0.5$

$$FC_f = \frac{4{,}288{,}548}{e^{(83)(0.35)}}$$

$$= < 1/100 \text{ ml}$$

   c. During final discharge period t = 130d, T = $10^\circ C$, $k_{FC} = 0.5$

$$FC_f = \frac{10{,}000{,}000}{e^{(130)(0.25)}}$$

$$= < 1/100 \text{ ml}$$

So: Disinfection will not be required,

5. Determine L_w based on soil permeability.

Use Equation (10-2) with 4% safety factor to determine design percolation rate (P_w):

$$P_w = (k)(24)(0.04)$$
$$= (0.5)(24)(0.04)$$
$$= 0.48 \text{ in/d}$$

$$\text{Annual } P_w = (0.48)(235 \text{ operating d}) = 112.8 \text{ in.}$$
$$= 9.4 \text{ ft/yr}$$

Use Equation (10-1) to determine hydraulic loading, L_w:

$$L_w = (ET-P_r) + P_w$$
$$= \frac{2}{12} + 9.4$$
$$= 9.6 \text{ ft/yr}$$

6. Determine L_{wn} based on nitrogen loading

Use Equation (10-15):

$$L_{wn} = \frac{(C_p)(P_r-ET) + 0.37\,(U-S_n)}{(1-f)(C_n) - C_p}$$

C_n = 19.7 mg/L (see step 3 above)

C_p = 10 mg/L (required by State)

f = 0.15

S_n = 0

U = 280 lb/acre·yr

$$L_{wn} = \frac{(10)(\frac{2}{12}) + 0.37\,(280)}{(0.85)(19.7)-10}$$
$$= 15.6 \text{ ft/yr}$$

7. Determine design hydraulic loading L_{wD}

L_w = 9.6 ft/yr (from Step 5)

L_{wn} = 15.6 ft/yr (from Step 6)

so:

$L_{wD} = L_w$ = 9.6 ft/yr

8. Determine field area

Use Equation (10-17):

$$A_f = \frac{(3.06)(Q)(V_s)}{L_{wD}}$$

$$Q = \frac{(100,000)(365)}{10^6} = 36.5 \text{ mg/yr}$$

$$V_s = 24,250 \text{ ft}^3/\text{yr} = 0.56 \text{ acre-ft/yr (from step 2)}$$

$$A_f = \frac{(3.06)(36.5) + 0.56}{9.6}$$

$$= 11.7 \text{ acres}$$

9. Check to see if grease content is LDP.

Grease = 0.2% = (20,000 mg/L)

$$\text{Annual loading} = \frac{(100,000)(365)(62.4)(.002)}{(7.48)(2000)} = 304 \frac{\text{ton}}{\text{yr}}$$

Assume none is removed in pond.

Then, maximum annual surface loading $= \frac{304}{11.7} = 26 \frac{\text{tons}}{\text{acre·yr}}$

See discussion in Chapter 3: Tolerant crops (perennial grasses) allow 3% of oil by weight in top 6 in. of soil. Assume soil = 100 lb/ft^3.

$$\text{Allowable load} = \frac{(100)(0.5 \text{ ft})(43,560)(0.03)}{2000} = 32.7 \frac{\text{tons}}{\text{acre·yr}}$$

So: Grease content is not LDP.

10. Check nutrient requirements for perennial grass cover. (see Table 5-8).

Nitrogen: 280 lb/acre·yr required
surplus available in wastewater

Phosphorus: 35 lb/acre·yr required
surplus available in wastewater

Potassium: Use Equation (5-2)

$$K_f = 0.9(U) - K_{ww}$$

$$K_{ww} = (2.7)(9 \text{ mg/L})(9.6 \text{ ft/yr})$$

$$= 233 \text{ lb/acre}\cdot\text{yr}$$

$$K_f = (.9)(280) - 233$$

$$= 19 \text{ lb/acre}\cdot\text{yr}$$

A supplemental potassium fertilization with 19 lb/acre·yr will be needed to maintain crop growth.

11. Estimate the BOD in the pond effluents (temperatures specified in step 3 above).

    a. After 12 h aeration. ($T = 10^\circ C$, $k_{20} = 2.5$)

(Equation 8-1)

$$C = \frac{(900)}{1 + (1.11)(\frac{12}{24})} = 578 \text{ mg/L}$$

    b. At end of 130 day storage period

(Equation 8-4)

$$\text{average } t = 65 \text{ d}$$

$$K_p @10^\circ C = 0.04$$

$$C = (58)e^{-(0.04)(65)}$$

$$= 42.9 \text{ mg/L}$$

    c. During first 83 days of pond discharge

$$C = (42.9)e^{-(0.04)(83)}$$

$= 1.6$ mg/L

d. During final 152 day discharge period

$$C = (578)e^{-0.04(152)}$$

$= 1.3$ mg/L

So there should be no odors at the application site due to BOD loading.

12. Determine system layout and schedule.

From Step 2: 145,950 ft^3/wk applied for a 33.6 wk application season

From Step 8: Field area = 11.7 acres

$$\text{Weekly hydraulic loading} = \frac{145{,}950}{(11.7)(43{,}560)} = 0.286 \text{ ft/wk}$$

$$= 3.4 \frac{\text{in.}}{\text{wk}}$$

Soil intake rate = 0.6 in./h

Use sprinklers with application rate = 0.4 in./h

Establish eight 1.7 acre fields

Total area = (8)(1.7) = 13.6 acres

This will allow for one "reserve" field.

$$\text{Sprinkler operation} = \frac{3.4 \text{ in./wk}}{0.4 \text{ in./h}} = 8.5 \frac{\text{h}}{\text{wk}}$$

Operate sprinklers on one field per day in rotation. Skip one field, in sequence, each cycle.

See Chapter 9 for details on sprinkler design and layout.

13. Groundwater aspects: If groundwater is always at the specified depth, significant mounding from the operation is unlikely. If the groundwater had been within 10 ft or so of the surface then the procedures in Chapter 4 should be used. Shallow seasonal groundwater would require either underdrains or termination of site operations during the period of saturation.

TOXIC AND HAZARDOUS WASTES

In 1983 there were about 200 land treatment sites receiving toxic or hazardous wastes (8)(9). Most of these were industrial operations with the majority (over 100) at petroleum refineries. Typically, the wastes are applied to the soil surface and mixed with the topsoil layer. When surface conditions permit, grass is then planted. The design waste loading and the number of repetitions is dependent on the factors discussed in Chapter 3. It might range from a single application of inorganic wastes to a number of periodic applications for organic materials such as solvents and oil wastes.

A treatment demonstration using the specific toxic or hazardous waste material with the expected site soils and operating conditions is essential for these types of waste and is required by 1983 EPA regulations. This demonstration will define the acceptability of particular wastes, the loading rates and the loading cycles for design. Other requirements for the final operational site include:

1. Five foot depth of unsaturated soil to function as the "treatment zone".
2. A three foot interval between the bottom of the

treatment zone and the seasonally high groundwater table.

3. Surface runoff and run-on controls designed for a 25 year storm. The runoff system must be able to collect and control a volume equal to the 24 hour, 25 year storm.
4. The groundwater must be monitored "upstream" and "downstream" of the application site.
5. The soil and soil moisture immediately beneath the treatment zone must be routinely monitored.

See Chapter 16 for additional information and criteria.

REFERENCES

1. U.S. DOI, Drainage Manual, U.S. GPO, No. 024-003-00117-1, Washington, D.C. (1978).
2. State of California Irrigation with Reclaimed Municipal Wastewater-A Guidance Manual, California State Water Resources Control Board, Sacramento, California (1984).
3. Holdridge, L.R., The Determination of Atmospheric Water Movements, Ecology 43:1-9, (1962).
4. Flach, K.W., Land Resources. In: Recycling Municipal Sludges and Effluents on Land. Champaign, University of Illinois, (July 1973).
5. State Department of Water Resources, Vegetated Water Use in California, California SDWR Bulletin 113-3 (1975).
6. Reed, S.C., et al, Wastewater Management by Disposal on the Land, U.S.A. CRREL SR 171, U.S.A. CRREL, Hanover, NH, (1972).
7. U.S. EPA, Process Design Manual for Land Application of Municipal Sludge. U.S. EPA CERI, Cincinnati, OH, (1983).
8. Morrison, A., Land Treatment of Hazardous Waste. ASCE Civil Engineering. 53(5): 33-38, (May 1983).
9. U.S. EPA, Hazardous Waste Land Treatment, SW-874, OSW, U.S. EPA, Washington, D.C., (April 1983).

11

Process Design—Overland Flow Systems

SYSTEM CONCEPT AND COMPONENTS

Overland flow (OF) is defined as the controlled application of wastewater on grass-covered, uniformly graded, gentle slopes, with relatively impermeable surface soils. The process was first applied in the United States for industrial wastewaters (1)(2)(3) and is still widely used for that purpose, particularly within the food processing industry. Early application of similar concepts for municipal wastewaters occurred in England where it is termed grass filtration, and in Melbourne, Australia (4). Many of these systems have been in continuous and successful operation since the late 19th century. Recent research efforts in the U.S. have focused on treatability of various wastes (5)(6)(7), performance at operational systems (8)(9)(10)(11) and on development of rational design criteria (12)(13)(14).

Site Characteristics

Overland flow is best suited for use at sites having surface soils that are slowly permeable or have a restrictive layer such as a claypan at depths of 1 to 2 ft. Overland flow can also be used on moderately permeable soils using higher loading rates than would be possible with an SR system. An OF system is also possible on very permeable soils by constructing an artificial barrier to prevent downward water movement through the soil, although the capital costs of such construction may be prohibitive for all but the smallest systems. If a significant portion ($>$20%) of the applied wastewater is expected to percolate and mingle with the groundwater then the nitrate considerations discussed in Chapter 3 may be a design factor for OF. When slowly permeable soils (percolate $<$20% of applied loading) are used for OF it is usually acceptable to disregard any groundwater impacts when municipal wastewaters are applied.

Overland flow may be used at sites with gently sloping terrain with grades in the range of 1 to 12%. Slopes can be constructed on nearly level terrain and terraced construction can be used when the natural slope grade exceeds 10%. In general, most of the operational systems constructed to date have grades ranging from 2 to 8 percent. At the wastewater application rates in typical use, the site grade, within the range of 2% - 8%, is not a critical factor in system design. Site grades less than 2% may require special measures to avoid ponding, and design grades greater than 8% increase the risk of short-circuiting and erosion.

System Configuration

The general system layout should match as closely as possible the natural topography at the site to avoid expensive earth moving during construction. The total area required for treatment is determined by methods described in this chapter. Individual slopes are then marked on the site plan until the total area required is satisfied. These individual slopes must also be connected with a network of ditches for collection of treated runoff and stormwater runoff for conveyance to the final system discharge point.

The system layout is also influenced by the type of wastewater distribution system selected. High solids content wastewaters typically use high pressure sprinklers to insure more uniform distribution of solids on the treatment slope. Low pressure systems involving gated pipe or sprinklers have been used successfully for screened, primary, secondary and pond effluents. Figure 11-1 illustrates the various possibilities for both low and high pressure types. Chapter 9 contains design details on both types of distribution systems.

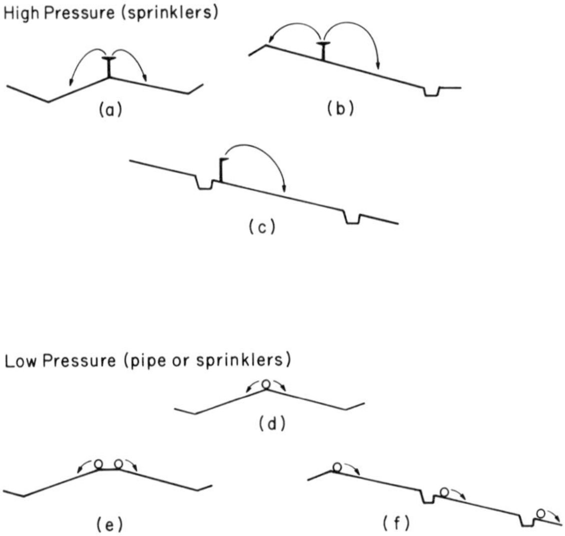

Figure 11-1 Distribution alternatives for overland flow, After (15).

Most of the early industrial systems were of the type shown in Figure 11-1 (a) or (b), with the sprinklers for type (b) located at about 1/3 the way down the slope so that all applied wastewater is contained on the treatment surface. Empirical criteria were developed through trial and error experience, so that slope lengths from 100 to 150 ft in length would provide adequate treatment for most wastewaters. If, for example, a sprinkler with a 100 ft diameter wetted circle is located at the 1/3 point on a 150 ft long slope, the "average" travel distance for all of the applied wastewater would then be 100 ft. If the solids content permits the use of low pressure systems, a gated pipe at the top of a 100 ft slope should therefore provide the same degree of treatment as the 150 ft slope with the pressure sprinkler at the "one-third" point. Low pressure systems are not suitable for high solids content wastewater since deposition of solids will occur in the immediate vicinity of the application point resulting in excess accumulation and odors.

Performance Standards and System Capabilities

Most overland flow systems have an outlet to surface waters and therefore require discharge permits. In the majority of cases the permit will limit BOD and TSS, and that is the basis for the design approach presented in this chapter. If the permit contains other requirements (ie: nitrification of ammonia, phosphorus removal, etc.) then Chapter 3 should be consulted to determine the limiting design parameter (LDP) for the system. The design procedure in these cases is a multi-step procedure:

1. Determine slope length, loading rates, etc. for BOD removal.
2. Determine, or estimate, slope length, etc., required for other parameters of concern.
3. The most conservative results of 1 and 2 above defines the LDP for the system and is the basis for system design.

The effluent quality from properly designed systems can consistently produce effluents with 10 mg/L of BOD and SS and with approximately 70% nitrogen removal (15). In concept, the system can be thought of as a plug flow, attached growth reactor with a vegetated surface. The near surface soil and surface deposits and the grass stems and roots provide the substrate for the biological components and the grass also serves as a sink for nutrients, as well as water removal via evapotranspiration.

Vegetation on the treatment slope is essential to regulate the flow and eliminate erosion, short-circuiting and channeling. The choice of vegetation is more limited for OF as compared to SR since perennial water tolerant grasses are the only feasible possibility for OF systems. These are discussed in detail in Chapter 5 and in the Construction section of this chapter. Chapter 14 on operation and maintenance also contains guidance for grass covered OF slopes. Rice, Reed canarygrass and similar grasses and reeds have the capability to transmit oxygen from the leaves through the stem to the root system. They can therefore flourish in a generally anaerobic environment and will in fact provide aerobic micro sites adjacent to the root filiments. The treatment responses observed in rice fields and highly loaded OF systems are generally similar. Reed canarygrass has become the vegetation of choice for many of the industrial OF systems for this reason.

In many respects OF systems offer greater control and flexibility in design and operation than the SR or RI concepts. In the typical SR or RI system there is no access to the wastewater once it is applied and penetrates the soil surface. All of the responses and constraints have to be anticipated and programmed into the design since there will be limited opportunities to control the responses once the system is operational. In contrast, most of the wastewater is continuously accessible in an OF system and this permits greater flexibility in operational adjustments. Since BOD is often the LDP for municipal systems, the engineer using the procedures in this chapter, can optimize the slope length required for a particular combination of wastewater quality and discharge requirements.

DESIGN PROCEDURES

The design approaches to be used for BOD, nitrogen, phosphorus and other LDP constituents are described below. In addition, the physical design is included since the system must insure uniform sheet flow of applied wastewater and have the capacity to convey stormwater runoff.

BOD

Laboratory and field research at the University of California-Davis (13)(14)(15) has resulted in the development and validation of a rational design procedure for overland flow when BOD is the LDP. The design model, based on first order, plug flow kinetics can be described with the following general form:

$$\frac{C_z - C}{C_o} = e^{-k\,(z/q^h)} \tag{11-1}$$

where C_o = concentration of BOD in applied wastewater, mg/L

C_z = concentration of BOD at distance (z) down the slope, mg/L

C = minimum achievable effluent BOD concentration (see discussion below) mg/L

A = empirically determined coefficient, dependent on q

z = actual distance down slope for gated pipe and other low pressure devices. The "average" travel distance for high pressure sprinklers, ft

k = empirical rate constant

q = wastewater application rate, $ft^3/h \cdot ft$ (the ft in the denominator is per foot of slope width)

n = empirical coefficient

The minimum achievable effluent concentration C, must be included since field work at several sites indicates that effluent from an OF slope will approach a non-zero, steady state value regardless of slope length. This "baseline" value is due to release of organic materials by the soil plant system. At the field system in Davis, CA used for model development, the value of C = 5 mg/L was determined (14).

The valid range for application rates (q) in Equation 11-1 is from 1 $ft^3/h \cdot ft$ to 4 $ft^3/h \cdot ft$. This upper limit exceeds the range previously used for design in the U.S. (2 $ft^3/h \cdot ft$). That earlier design approach expressed hydraulic loadings for OF in terms of a depth of water over the entire treatment area in a manner similar to that used for SR systems (Chapter 10). The example below is presented to clarify the relationship between the two parameters.

Example 11-1: Determine the Wastewater Application Rate for an Operational OF System

Conditions: An operational OF system has slopes 120 ft long and 100 ft wide with gated pipe at the top. The average weekly application is 9 in. of wastewater. Wastewater is applied 8 h/d for 5 d/wk. Find the application rate q in $ft^3/h \cdot ft$.

Solution:

$$\text{Treatment area} = (120)(100) = 12{,}000\ ft^2$$

$$\text{Wastewater Applied} = \left(\frac{9\ in/wk}{12\ in/ft}\right)(12{,}000\ ft^2) = 9000\ \frac{ft^3}{wk}$$

$$\text{Application rate, } q = \left(\frac{9000\ ft^3}{wk}\right)\left(\frac{1}{8\ \frac{h}{d}}\right)\left(\frac{1}{5\ \frac{d}{wk}}\right)\left(\frac{1}{100\ ft}\right)$$

$$= 2.25\ ft^3/h \cdot ft$$

Equation 11-1 can be solved for either z or q depending on site conditions. The existing topography on portions of the site may limit the available slope length, so the equation is then used to determine q for those areas. Where there is no constraint on potential slope length a q is assumed and z calculated. It is desirable to have as many identical slopes as possible in the treatment system to reduce costs and complexity of conveyance and pumping. This is easy to achieve when a system is constructed on essentially flat ground by cut and fill. When the existing site relief is to be utilized in the system, slope lengths for a range of application rates should be calculated. The site topographic map (plotted to 1 ft contours) should then be examined to determine which of the calculated slope lengths can be used most advantageously for the site. That length is then used for layout of all slopes until sufficient area is provided.

Rearranging Equation (11-1) to solve for slope length produces:

$$z = \frac{\ln\left[\frac{C_z - C}{AC_o}\right]}{(-K)} \qquad (11\text{-}2)$$

where K = overall rate constant

$= \frac{k}{q^n}$

other terms as defined previously

Both The K and the A factors are dependent on the application rate (q) and on the type of wastewater being applied. They can be estimated with the following equations for the conditions specified:

Screened Raw Wastewater, $q = 1$ to 4 $ft^3/h \cdot ft$

BOD removal : $A = 0.052(q + 7.78)$ (11-3)

$K = 0.003\ (8.67 - q)$ (11-4)

Primary Effluent, $q = 1$ to 4 $ft^3/h \cdot ft$

BOD removal : $A = 0.125\ (q + 1.54)$ (11-5)

$K = 0.001\ (18.6 - q)$ (11-6)

all terms defined previously

Reference 14 contains a more exact graphical solution for Equation 11-2 but Equations 11-3 through 11-6 will provide the basis for a reasonable approximation. The A and K factors were evaluated for screened raw sewage and primary effluents as noted above. The latter equations (11-5 and 11-6) can also be used to provide a conservative estimate of slope length required if secondary effluent is applied.

Example 11-2: Find the Required Slope Length for Various Levels of Preapplication Treatment

Conditions: Assume an industrial wastewater with BOD = 700 mg/L after screening, BOD = 350 mg/L after primary, BOD = 30 mg/L after secondary. Discharge requirement = 10 mg/L BOD, assume application rate of 0.25 gpm/linear ft of slope width with application 10 h/d, 6 d/wk. High solids content wastewater, so high pressure sprinklers required.

Find: "Average" travel distance for each type of wastewater. This will permit cost comparison of OF site development versus costs for higher levels of preapplication treatment.

Solution:

$$\text{Application rate } q = \frac{(0.25 \text{ gpm/ft})(60 \frac{\text{min}}{\text{h}})}{(7.48 \text{ ft}^3/\text{gal})} = 2.0 \frac{\text{ft}^3}{\text{h}\cdot\text{ft}}$$

1. Case 1 Screened wastewater:

$$\frac{C_z - C}{C_o} = \frac{(10-5)}{700} = 0.0071$$

$$A = 0.052\ (q + 7.78)$$

$$= 0.052\ (2.00 + 7.78)$$

$$= 0.508$$

$$K = 0.003\ (8.67 - q)$$

$$= 0.020$$

$$z = \frac{\ln(\frac{0.0071}{0.508})}{(-0.020)} = 214 \text{ ft}$$

2. Case 2 Primary effluent:

$$\frac{C_z - C}{C_o} = \frac{5}{350} = 0.0143$$

$$A = (0.125)(q + 1.54) = 0.442$$

$$K = (0.001)(18.6 - q) = 0.0167$$

$$z = \frac{\ln\left(\frac{0.0143}{0.442}\right)}{(-0.0167)} = 205 \text{ ft}$$

3. Case 3 Secondary effluent:

$$\frac{C_z - C}{C_o} = \frac{5}{30} = 0.1667$$

A = 0.442 (previous case)

K = 0.0167 (previous case)

$$z = \frac{\ln\left(\frac{0.1667}{0.442}\right)}{(-0.0167)} = 58 \text{ ft}$$

4. The difference in slope length between case 1 and case 2 is not significant so there is no purpose in providing primary treatment as compared to applying screened wastewater. The cost analysis would then compare the additional 156 ft (214-58) of treatment slope to the costs of a complete secondary treatment plant. It is likely that overland flow would be more economical.

5. A 214 ft "average" travel distance, with the high pressure sprinkler at the 1/3 point on the slope would require a total slope length of 320 ft. Assuming each treatment slope was 214 ft wide, the weekly hydraulic loading on the wetted area (assuming multiple sprinklers for uniform coverage) would be:

$$(2.0 \text{ ft}^3/\text{h}\cdot\text{ft})(214 \text{ ft})(10 \tfrac{\text{h}}{\text{d}})(6 \tfrac{\text{d}}{\text{wk}}) = 25{,}680 \text{ ft}^3/\text{wk}$$

$$\text{Treatment area} = (320)(214) = 68{,}480 \text{ ft}^2$$

$$\text{Hydraulic loading} = \frac{25{,}680}{68{,}480} = 0.375 \frac{\text{ft}}{\text{wk}} = 4.5 \frac{\text{in.}}{\text{wk}}$$

6. If the wastewater had a low solids content (<80 mg/L) gated pipe or other forms of low pressure distribution could be used and the slope length would only need to be 214 ft. Assuming the same width as the case above the hydraulic loading would then be:

$$\text{Treatment area} = (214)(214) = 45{,}796 \text{ ft}^2$$

$$\text{Hydraulic loading (for gated pipe)} = \frac{25{,}680}{45{,}796} = 0.56 \frac{\text{ft}}{\text{wk}} = 6.7 \text{ in./wk}$$

Suspended Solids

With the exception of algae, wastewater solids will not be the LDP for overland flow design. Suspended and colloidal solids are effectively removed because of the low velocity and the shallow depth of flow on the treatment slope. Maintenance of a thick grass cover and elimination of channel flow are essential for solids removal. The removal of suspended matter is relatively unaffected by cold weather (17) or other process loading parameters.

When lagoons or storage ponds are used in overland flow systems the presence of algae in the wastewater may result in high suspended solids in the final slope effluent due to the inability to remove some types of algae (18). Many small diameter, free floating species of algae and diatoms have little or no tendency to aggregate and are particularly difficult to remove. Examples are green algae *Chamdomonas* and *Chlorella* and the diatoms *Anomoeoneis*. In contrast, the green algae *Protococcus* has a "sticky" surface and is effectively removed on the OF slope. Since control of algal species in the lagoons or ponds is not a practical possibility it is necessary to by-pass these components during periods of algal bloom to sustain high quality OF effluent when stringent SS limits are required.

Nitrogen

The removal pathways, discussed in detail in Chapters 3 and 5, include crop uptake, ammonia volatilization and nitrification/denitrification. As indicated previously about 70% of the applied nitrogen is removed on the OF slope with most municipal wastewaters within the range of application rates indicated in the BOD section. About 20% of this removal is due to crop uptake (with harvest and removal of the crop), ammonia volatilization will account for about 10% and nitrification/denitrification for the balance. If the crop is cut but not removed, full removal credit cannot be taken for all of the nitrogen contained. The mineralization rates discussed in Chapter 3 should be used to estimate the amount of nitrogen released each year by the decomposing grass.

As with other attached growth processes, nitrification is quite rapid and very effective on the OF slope as long as suitable conditions prevail. Temperatures in the range of 50° - 70° F for the wastewater and the soil surface are necessary for effective nitrification. Soil temperatures below 45° F will tend to inhibit nitrification. An indication of the relationship was developed with field research in New Hampshire (17):

$$E_{NH} = 0.114\,(T)^2 - 3.94\,(T) + 35.1 \qquad (11\text{-}7)$$

where E_{NH} = effluent ammonia concentration, mg/L

T = soil temperature at a depth of 1 in., °C

This equation was developed for an application rate of about 0.5 $ft^3/h \cdot ft$. Another equation, relating ammonia to detention time on the slope was developed from data at the same facility:

$$E_{NH} = (0.81)(I_{NH})e^{-1.8t} \qquad (11\text{-}8)$$

where E_{NH} = effluent ammonia concentration, mg/L

I_{NH} = ammonia concentration in applied wastewater, mg/L

t = detention time, h

Neither Equation (11-7) or (11-8) have been validated at other sites but they should provide a reasonable estimate for preliminary designs. Both nitrification and denitrification on the slope are favored by the least possible level of preapplication treatment. This will leave the wastewater nitrogen in the organic and ammonia forms which are easier than nitrate to retain and will also supply the additional organic carbon needed for denitrification.

Another approach, based on the field work described in reference 14 relates ammonia remaining in the wastewater to the down slope distance:

Screened raw wastewater:

$$E_{NHz} = 36-0.145z \qquad (11\text{-}9)$$

Primary effluent:

$$E_{NHz} = 26-0.128z \qquad (11\text{-}10)$$

where E_{NHz} = ammonia concentration at point z, mg/L

z = distance down slope, ft

Equation (11-9) and (11-10) should provide a reasonable estimate for application rates (q) of 2 $ft^3/h\cdot ft$ or less. At higher application rates and with continuous 24 hour per day operations nitrification is less effective. Reference 14 can be used to graphically estimate nitrogen removal at $q > 2\ ft^3/h\cdot ft$.

A design based on BOD removal, for typical municipal effluents, will provide enough treatment surface to obtain the typical 70% nitrogen removal. If very stringent discharge standards or high wastewater nitrogen content (> 60 mg/L) are factors, then the surface area required for N removal should be determined with the procedures described in Chapter 10 (with appropriate modifications, ie: no percolate nitrogen) and compared to the area required for BOD removal. The mass balance Equation (10-9) becomes:

$$L_{OFN} = U + D + 2.7\ E_N\ Q_A \qquad (11\text{-}11)$$

where L_{OFN} = mass loading of nitrogen on overland flow area, lb/acre·yr

U = crop uptake lb/acre·yr

D = site losses:volatilization, denitrification, lb/acre·yr

where E_N = allowable effluent nitrogen concentration, mg/L

Q_A = annual system runoff, ft/yr

Phosphorus

As discussed in Chapter 3, the slight contact between the soil and the wastewater in OF systems limits phosphorus removals to about 50%. Extending the slope length or reducing the application rate will not be effective over the long run. If high P removals are required then application of alum at a 1:1 ratio with P will precipitate most of the phosphorus either on the slope or in a preapplication step.

Other Constituents

In general, materials such as trace elements, trace organics, bacteria and virus will not be the LDP for design of overland flow. See chapter 3 for a discussion of the fate of these materials. Overland flow effluent will be comparable to undisinfected conventional secondary treatment effluent with respect to bacteria and virus.

LAND AREA REQUIREMENTS

The total treatment area required is based on the application rate assumed and the slope length calculated for the LDP. The LDP in some cases may be nitrogen but in general it will be BOD and the slope length is determined with Equation (11-2). The total area can then be calculated with the following equation:

$$A_F = \frac{(Q_D)(z)(7)}{(q)(t)(d)(43,560)} \qquad (11\text{-}12)$$

where A_F = field area for treatment, acres

Q_D = design daily flow, ft^3/d

z = actual slope length [Equation (11-2) for low pressure distribution; Equation (11-2) plus adjustment for high pressure distribution]

q = application rate, $ft^3/h \cdot ft$

t = daily application period, h/d

d = number of operating days per week, d/wk

Example 11-3: Determine Treatment Area for the System Described in Example 11-2

Conditions: $q = 2.0\ ft^3/h \cdot ft$, $z = 320$ ft (adjusted for high pressure sprinklers), $t = 10$ h/d, $d = 6$ d/wk, 500,000 gpd design flow.

Solution:

1. $$Q_D = \frac{500,000}{7.48} = 66,845\ ft^3/d$$

$$A_F = \frac{(66,845)(320)(7)}{(2)(10)(6)(43,560)}$$

$$= 28.6 \text{ acres}$$

2. If gated pipe could have been used, then $z = 214$ ft, and:

$$A_F = 19 \text{ acres}$$

In addition to the treatment area calculated with Equation 11-12, land must be provided for roads, O&M buildings, preapplication treatment, unusable portions of the site and possibly buffer zones. At sites with extensive surface relief the unusable portions of the site may be higher for an OF system than for SR. At existing OF systems, the area provided for these other purposes ranges from 15 to 40% of the treatment area (not including buffers or wastewater storage).

Application Period

A wide range of application periods has been used successfully, ranging from just a few hours to as high as 24 h/d. The application periods that have been used most frequently in existing OF projects range between 6 and 12 h/d. Use of design application periods of 12 h/d or less allows more operating flexibility during periods when parts of the system must be shut down for harvest or repair.

Application Schedule

A design application frequency of 7 d/wk is generally used to minimize land area requirements and eliminate or reduce storage requirements. There does not appear to be any advantage in terms of process performance to using less frequent applications. For small systems with storage facilities, it may be more convenient to use an application frequency of 5 d/wk and shut down the OF on weekends.

Winter Operation vs Storage

Due to the limited operating experience with OF in different parts of the country, cold weather storage requirements are not well defined. In general, OF systems must be shut down for the winter when effluent quality requirements cannot be met due to cold temperatures even at reduced application rates or when ice begins to form on the slope. The duration of the shutdown period and, consequently the required storage period will, of course, vary with the local climate and the required effluent quality. The most conservative approach would assume a storage period equal to that required for SR systems (Chapters 8 and 10). At wastewater/soil temperatures above 50° F the BOD removal efficiency is independent of temperature (14). Low temperature studies in New Hampshire (17) indicated the following relationship:

$$E_{BOD} = (0.226)(T)^2 - 6.53(T) + 53 \qquad (11\text{-}13)$$

where E_{BOD} = effluent BOD, mg/L

T = soil temperature, °C

This equation was developed for a loading rate of about 0.5 $ft^3/h \cdot ft$. At a soil temperature of less than 39° F the effluent BOD will exceed 30 mg/L. A pilot scale OF system was operated continuously over two winters in Harriman, NY (19) and consistently achieved at least 30/30 effluent quality for both BOD and SS when either primary or secondary effluent was applied. The site, in the Hudson River valley experienced below freezing ambient air temperatures from early December through late January. It is probable that the warm wastewater and the frequent applications were sufficient to maintain acceptable conditions at the treatment surface and significant ice did not accumulate or persist on the slopes.

The presence of standing grass and a snow cover will provide significant insulation and possibly maintain adequate temperatures in the treatment zone. It is necessary to terminate operations when ice forms and remains on the slope. Runoff in these cases will likely occur on top of the ice sheet with little treatment provided.

An alternative to winter storage is to design the treatment slopes for a lower winter application rate to insure acceptable performance.

Table 11-1 summarizes criteria from the EPA process design manual for such cases (20).

Table 11.1: U.S. EPA Criteria for Winter Operation of Overland Flow (20)

Level of Pretreatment	q ($ft^3/h{\cdot}ft$)	t (h/d)	d (d/wk)
Screened	0.75	8-12	7
Primary	0.86	8-12	7
Activated sludge or similar	1.08	8-12	7
Lagoons	0.32	8-12	7

Two variations of this alternative are possible:

(a) Size the slopes for winter application rates with storage for emergencies and ice periods only.

(b) Size treatment slopes for higher summer rates. Apply wastewater in winter at winter rates and store the balance.

A cost comparison will indicate which alternative is the most effective. If storage is needed for other purposes then alternative (b) will probably be the most economical. The total treatment area required for this second alternative can be determined with:

$$A_F = \frac{\left[Q_D(365) + V_s\right] z}{\left[(q_w)(t_w)(d_w) + (q_s)(t_s)(d_s)\right] (43,560)} \tag{11-14}$$

where A_F = field area, acres

Q_D = annual average daily flow in community or industry, ft^3/d

V_s = net gain or loss of water from storage, ft^3/yr (precip. or evaporation)

q_w = winter application rate, $ft^3/h \cdot ft$

t_w = winter operational time, h/d

d_w = winter schedule, d/wk

q_s = spring, summer, fall application rate, $ft^3/h \cdot ft$

t_s = spring, summer, fall operational time h/d

d_s = spring, summer, fall schedule, d/wk

The duration of the reduced loading period, q_w, is about 90 days at existing systems. If storage is provided for winter periods, the storage pond should be bypassed in the warm months to minimize algal problems. The storage pond contents can then be blended with the bypassed wastewater for application on the slopes.

Storage for Rainfall Runoff

Research and field studies at a number of systems (2)(14)(20) indicate that rainfall runoff either during or after wastewater applications did not significantly affect the concentration of the major constituents in the runoff. However, due to the increased flow, the mass of constituents discharged does increase.

Recent work at the Davis, CA system indicates that these discharges are due to losses of natural organics and litter on the slope and not wastewater constituents and in fact were less than the losses from control slopes where no wastewater had been applied. When mass discharges are the controlling parameter for permits, it is necessary to obtain higher discharge allowances during storm events or during high flow periods in the receiving stream. The alternatives are to collect and recycle part of the stormwater runoff or to store it until it attains acceptable quality for discharge. Reference 20 provides a detailed procedure for determining storage requirements in these cases.

PHYSICAL FEATURES AND CONSTRUCTION DETAILS

Runoff Collection

The collection channels are usually vegetated with the same species of grasses growing on the slopes and should be graded to prevent erosion. There are some cases, however, where additional construction is necessary. Sharp bends or steep grades along runoff channels will increase the potential for erosion, and it may be necessary to provide additional protection in the form of riprap, concrete, or other stabilizing agents at these points. Runoff channels should be graded to no greater than 25% of the slope grade to prevent unwanted cross flow on the slope.

In humid regions, particularly where the topography is quite flat and the runoff channels have small grades, grass covered channels may not dry out entirely. This may increase channel maintenance problems and encourage mosquito populations. In these cases, concrete or asphalt can be used or a more elaborate system involving porous drainage pipe lying in the channel beneath a gravel cover. Small channels are normally V-shaped, while major conveyance channels have trapezoidal cross-sections. Backwater will occur on the slopes and create marshy areas unless the ditch has sufficient capacity and elevation drop to convey the design volume. Lined ditches or perforated pipe have a higher initial cost but may be easier to maintain than an unlined ditch. The top edge of the ditch lining in these cases should be placed at least 8 in. below the treatment slope surface to allow drainage of the surface soil layer (15).

In addition to transporting treated effluent to the final discharge point, the runoff channels must also be capable of transporting all stormwater runoff from the slopes. The channels should be designed, at a minimum, to carry runoff from a storm with a 25 year return frequency. Both intensity and duration of the storm must be considered. A frequency analysis of rainfall intensity must be performed and a rainfall-runoff relationship developed to estimate the flow rate due to storm runoff that must be carried in the channels. The local SCS office can provide assistance in performing this design. In some cases, it is desirable to provide a perimeter drainage channel around the OF site to exclude offsite stormwater from entering the OF drainage channels.

Compound or cross slope grades will minimize earth work and the depth of collection ditches. Cross slope grading results in a lateral component to the down slope flow. On square or long narrow slopes this will leave a significant portion of the surface area dry. On wide slopes the dry area will not be as significant. Such cross slope grades should not exceed 0.5%.

Distribution Systems

See Chapter 9 for detailed discussion of OF distribution, both low and high pressure systems. In practice the gated aluminum pipe commonly used for agricultural irrigation is suitable for use on OF systems with any type of wastewater. Adjustable plastic slide gates spaced on 2 ft centers are preferred. The gate opening are manually adjusted but at the application rates typically used in OF are open only a small portion of the total available. Routine inspection and cleaning are essential since fibrous material will accumulate at the gate and will occur regardless of the level of pretreatment used. Pipe lengths greater than 300 ft require in-line control valves as described in Chapter 9.

Access to pipes and/or sprinklers is essential for the operator and to avoid damage to the wet treatment areas. A grass free strip 4-6 ft wide is suggested, and gravel is the most commonly used surface treatment for these areas.

Site Grading

Smooth sheet flow down the slope is critical to consistent process performance, so emphasis must be placed on the proper construction of the slopes. Naturally occurring slopes, even if they are within the required length and grade range, seldom have the uniform overall smoothness required to prevent channeling, short-circuiting, and ponding. Therefore, it is necessary to completely clear the site of all vegetation and to regrade it into a series of OF slopes and runoff collection channels. The first phase of the grading operation should be accomplished within a grade tolerance of 0.1 ft. If a buried distribution system is being used, this grading phase is generally followed by the installation of the distribution piping and appurtenances.

After the slopes have been formed in the first grading operation, a farm disk should be used to break up the clods, and the soil should then be smoothed with a land plane. Usually, a grade tolerance of plus or minus 0.05 ft can be achieved with three passes of the land plane. Surface distribution piping may be installed at this stage.

Soil samples of the regraded site should be taken and analyzed by an agricultural laboratory to determine the amounts of lime and fertilizer that are needed. The appropriate quantities should then be added prior to seeding. A light disk should be used to eliminate any wheel tracks on the slopes as final preparation for seeding.

Vegetation

The various grass mixtures used for overland flow systems are described in Chapter 5. In the northern humid zones various combinations of orchard grass, Reed canarygrass, tall fescue and Kentucky bluegrass have been most successful. The use of a "nurse" grass such as perennial ryegrass is recommended since it will grow very quickly and protect the soil surface while the other grasses are becoming established. Placement of sod will provide an "instant" grass cover but will not be economical for most systems.

It has been found that a Brillion seeder is capable of doing an excellent job of seeding the slopes. The Brillion seeder carries a precision device to drop seeds between cultipacker-type rollers so that the seeds are firmed into shallow depressions, allowing for quick germination and protection against erosion. Hydroseeding may also be used if the range of the distributor is sufficient to provide coverage of the slopes so that the vehicle does not have to travel on the slopes. When seeding is completed, regardless of the means, there should be no wheel tracks on the slopes. Traffic on the slope in the direction of water flow should be avoided whenever possible to avoid future channeled flow. Vehicle access should be in the cross slope direction and only when the soil is dry.

A good vegetative cover is essential prior to application of wastewater. Grass planting should only be undertaken during the optimum periods for that activity in a particular location and the overall construction schedule must be adjusted accordingly. In arid and semi arid climates portable sprinklers may be necessary to provide moisture for germination and growth of the grass. The wastewater distribution system should not be used until the grass is established to avoid erosion of the bare soil. The construction contract should have a contingency to cover reseeding or erosion repair on the case of intense rainfall during the period between final site grading and grass establishment.

As a general rule, wastewater should not be applied at design rates until the crop has grown enough to receive one cutting. Cut grass from the first cutting may be left on the slope to help build an organic mat as long as the clippings are short (<1 ft); long clippings tend to remain on top of the cut grass thus shading the surface and retarding regrowth.

A period of slope aging or acclimation is required following initial startup before process performance approaches satisfactory levels. During this period, the microbial population on the slopes is increasing and slime layers are forming. The initial acclimation period may be as long as 3 to 4 months. If a variance to allow discharge during this period can not be obtained, provisions should be made to store and/or recycle the effluent until effluent quality improves to the required level.

An acclimation period also should be provided following winter storage periods for those systems in cold climates. Acclimation following winter shutdown should require less than 1 month. Acclimation is not necessary following shutdown for harvest unless the harvest period is extended to more than 2 or 3 weeks due to inclement weather. See Chapter 14 for additional details on operation and maintenance of OF systems.

COMBINATION SYSTEMS

When very stringent water quality standards prevail, and suitable soils are available, it may be cost effective to use OF as a component in a combined system, particularly with RI. This may be attractive for example when both nitrogen and phosphorus are an LDP. Overland flow as described previously is very effective at nitrogen removal and nitrification, less so at phosphorus removal because of the limited soil contact. RI systems, depending on the soil characteristics, tend to have the opposite capabilities. A combination of the two should then achieve high removals for both N and P. The sequence of the OF and RI components will depend on the general site conditions (topography, soils, groundwater location, etc.). The following example illustrates some of the possibilities.

Example 11-4: Evaluate the Feasibility of an Overland Flow-Rapid Infiltration Combination.

Conditions: As described in Examples (11-2) and (11-3), 500,000 gpd industrial flow. Discharge requirements: BOD/SS 5 mg/L, N = 10 mg/L, P = 0.5 mg/L. Raw wastewater: BOD = 700 mg/L, N = 60 mg/L, P = 20 mg/L. Site soils sandy loam: K = 1.0 in/h.

Solution: From Example 11-2 and 11-3.

1. Overland flow treatment area = 28.6 acres
2. Water quality of treated OF runoff.

	BOD/SS	= 10 mg/L
(70% removal)	N	= 18 mg/L
(50% removal)	P	= 10 mg/L

3. See Chapter 4, site soils are too permeable as-is for overland flow. So, strip off 8 - 12 in. topsoil, mix clay with subgrade and compact, after grading. Then replace topsoil and complete construction of OF slopes.

4. See Figure 4-6, site soils are moderately permeable. Acceptable RI hydraulic loading might be about 60 ft of wastewater per year. Assuming no water losses on OF slope:

$$\text{RI basin area} = \frac{(500{,}000)(365)}{(7.48)(60)(43{,}560)} = 9.3 \text{ acres}$$

Based on criteria in Chapter 12 percolate quality beneath the basin would be:

$$\text{BOD/SS} \leqslant 5 \text{ mg/L}$$

$$\text{Total N} \cong 9 \text{ mg/L}$$

P ≅ 0.5 mg/L after 1500 ft of vertical and lateral travel away from basin surface

These RI basins would have sufficient area to treat the wastewater flow. Depending on State requirements it might also be necessary to provide polishing ponds for OF storm water runoff.

5. If the groundwater table were relatively shallow (20 - 30 ft) it might be feasible to adopt the following sequence:

a. Primary treatment at industrial plant to reduce solids.

b. RI basins, loaded with primary effluent, for first assumption use: 15 acres.

c. Recovery wells and pumps 1,000 ft down-gradient from RI basins.

d. Quality of recovered water:

BOD/SS = < 10 mg/L

N = 30 mg/L

P = 1.0 mg/L

e. See Example 11-2, slope length for high quality water = 58 ft. Use gated pipe since low solids water. Overland flow treatment area:

$$A_f = \frac{(66,845)(58)(7)}{(2)(10)(6)(43,560)}$$

$$= 5.2 \text{ acres}$$

f. Water quality of OF runoff:

BOD/SS = 5 mg/L

N = 9 mg/L

P = 0.5 mg/L

g. Stormwater runoff from this OF system might also have to be contained but the contributing area in this case is much lower.

REFERENCES

1. Thornthwaite, C.W., An Evaluation of Cannery Waste Disposal by Overland Flow Spray Irrigation. Publ. Climatol. 22 (Sept. 1969).

2. Hinrichs, D.J., et al., Assessment of Current Information on Overland Flow of Municipal Wastewater, U.S. EPA 430/9-80-002, U.S. EPA, OWPO, Washington, D.C., (May 1980).

3. Gilde, L.C., et al., A Spray Irrigation System for Treatment of Cannery Wastes. Journal WPCF. 43: 2017 (1971).

4. Seabrook, B.L., Land Application of Wastewater in Australia. U.S. EPA 430/9-75-017, U.S. EPA, OWPO, Washington, D.C., (May 1975).

5. Thomas, R.E., et al., Overland Flow Treatment of Raw Wastewater with Enhanced Phosphorus Removal, U.S. EPA, Office of Research and Development. EPA-660/2-76-131, (1976).

6. Peters, R.E., et al., Field Investigations of Advanced Treatment of Municipal Wastewater by Overland Flow. Volume 2. Proceedings of the International Symposium on Land Treatment of Wastewater. Hanover, New Hampshire (August 1978).

7. Carlson, C.A., et al., Overland Flow Treatment of Wastewater U.S.A. WES Misc. Paper Y-74-3, U.S.A. WES, Vicksburg, MS., (Aug. 1974).

8. Scott, T.M. and Fulton, D.M., Removal of Pollutants in the Overland Flow (Grass Filtration) System. Progress in Water Technology, II, (4 and 5): 301-313 (1979).

9. Peters, R.E., Lee, C.R., and Bates, D.J., Field Investigations of Overland Flow Treatment of Municipal Lagoon Effluent. U.S.A. WES, Tech. Rpt., EL-81-9, U.S.A. WES, Vicksburg, MS., (Sept. 1981).

10. Hall, O.H., et al., Municipal Wastewater Treatment by the Overland Flow Method of Land Application. U.S.EPA 600/2-79-178, U.S.EPA RSKERL, Ada,OK., (August 1979).

11. Ketchum, L.H., et al., Overland Flow Treatment of Poultry Processing Wastewater in Cold Climates. U.S.EPA 600/S1-81-234, U.S.EPA RSKERL, Ada, OK. ,(September 1981).

12. Jenkins, T.F., et al., Pilot Scale Study of Overland Flow Land Treatment in Cold Climates. In: Proceedings: Developments in Land Methods of Wastewater Treatment and Utilization-Melbourne, Australia October 1978. Pergamon Press, New York. Progress in Water Technology, 11 (4-5): 207.

13. Smith, R.G., Development of a Rational Basis for Design and Operation of the Overland Flow Process. Proceedings: National Seminar on Overland Flow Technology for Municipal Wastewater, EPA 600/9-81-022. U.S.EPA, OWPO, Washington, D.C., (1981).

14. Smith, R.G., Schroeder, E.D., Demonstration of the Overland Flow Process for the Treatment of Municipal Wastewater-Phase II Field Studies, Dept. Civil Engr. , University of Calif.-Davis. Report to California State Water Resources Control Board., (1982).

15. Smith, R.G., Schroeder, E.D., Physical Design of Overland Flow Systems, Journal WPCF, 55(3): 255-260 (1983).

16. Witherow, J.L., Wastewater Irrigation of Rice, U.S. EPA 430/9-80-005. U.S. EPA, OWPO, Washington, D.C., (May 1980).

17. Jenkins, T.F., et al., Performance of Overland Flow Land Treatment in Cold Climates: In: Proceedings State of Knowledge in Land Treatment of Wastewater, Hanover, NH, Vol. 2, p. 61, U.S.A. CRREL, Hanover, NH, (August 1978).

18. Witherow, J.L., Bledsoe, B.E., Algae Removal by the Overland Flow Process, Journal WPCF, 55 (10): 1256-1262 (1983).

19. Clark, P.J., Marsh Pond/Overland Flow Pilot Plant Project Report, P.J. Clark Engineers, Rochester, NY, (Feb. 1983).

20. U.S. EPA, Process Design Manual for Land Treatment of Municipal Wastewater, EPA 625/1-81-013, U.S. EPA, CERI, Cincinnati, OH., (October 1981).

21. Palazzo, A.J., Plant Growth and Management for Wastewater Treatment in Overland Flow Systems, SR82-5, U.S.A. CRREL, Hanover, NH, (April 1982).

12

Process Design—Rapid Infiltration Systems

The process design of rapid infiltration systems is generally governed by the infiltration rate and permeability of the soil. Selection of the hydraulic loading rate can also affect the removal of nitrogen and phosphorus.

The preapplication treatment for RI systems ranges from primary treatment to secondary treatment (see Chapter 8). Hydraulic loading rates range from 20 to 400 ft/yr. As shown in Table 12-1 several RI systems have been operating for more than 40 years with application of primary effluent. The system at Whittier Narrows recharges the potable groundwater and is in an urban area, thus the preapplication treatment is tertiary (filtered secondary).

Table 12.1: Selected Rapid Infiltration Systems (1)-(10)

Location	Type of Wastewater	Year Started	Flow (mgd)	Hydraulic Loading (ft/yr)	Effluent Disposition
Calumet, MI	Raw	1887	1.6	116	Surface
Fontana, CA	Primary	1953	2.9	57	Groundwater
Ft. Devens, MA	Primary	1941	1.0	100	Surface
Hollister, CA	Primary	1946	1.0	50	Surface
Lake George, NY	Secondary	1939	1.1	140	Surface
Milton, WI	Secondary	1937	0.3	360	Groundwater
Phoenix, AZ	Secondary	1974	13.0	200	Reuse
Seabrook Farms, NJ	Screened cannery	1950	3.4	53	Groundwater
Vineland, NJ	Primary	1927	4.1	70	Surface
Whittier Narrows, CA	Tertiary	1963	12.5	160	Groundwater

Most of the 320 RI systems in the U.S. discharge the treated water indirectly into nearby surface water as shown in Figure 2-2.

The typical procedure for design of RI basins is as follows:

1. Determine the design infiltration rate (see Chapter 7).
2. Determine the RI hydraulic pathway, based on the site hydrogeology and discharge requirements to surface or groundwater.
3. Determine the treatment needs by comparing wastewater characteristics to the water quality requirements.
4. Select the preapplication treatment level appropriate for the site and the treatment needs (see Chapter 8).
5. Calculate the hydraulic loading rate based on the treatment needs, the infiltration rate and the preliminary wet/dry ratio.
6. Calculate the land requirements.
7. Check the potential for groundwater mounding and determine the need for underdrains (see Chapter 4).
8. Select a hydraulic loading cycle and the number of basin sets.
9. Calculate the application rate and check the final wet/dry ratio.
10. Layout the basins and design berms, structures, etc.
11. Determine monitoring requirements and locate monitoring wells (see Chapter 14).

TREATMENT REQUIREMENTS

The treatment performance at RI systems is relatively independent of infiltration rate for most constituents (see Chapter 3). For nitrogen, and to some extent phosphorus, the infiltration rate can affect the treatment performance.

Nitrification

As indicated in Chapter 3 for RI systems, application rates

of up to 12 in./d with 20 mg/L of ammonia will result in a nitrified effluent. As wastewater temperatures drop, the rate of nitrification will also decrease. For example, at temperatures of 40° to 45° F nitrification rates will be substantially less than rates at 70° F (11). Recent experience at Boulder, Colorado has shown that even though nitrification declines, at temperatures of 40° F there was still removal of ammonia to 1 mg/L or less from about 9 mg/L (12). Reducing application rates during cold weather will allow for these reduced nitrification rates and will also allow more of the applied ammonia to be adsorbed in the soil profile. For nitrification, the loading cycle should consist of short (1 day or so) application periods and relatively long (5 to 10 days) drying periods.

Nitrogen Removal

Nitrogen removal by denitrification requires both adequate organic carbon and adequate detention time. The potential carbon limitation on the amount of nitrogen removal can be approximated using the following equation:

$$N = \frac{TOC-5}{2} \qquad (12\text{-}1)$$

where N = change in total nitrogen, mg/L

TOC = total organic carbon in the applied wastewater, mg/L

The 5 mg/L of residual TOC is typical for municipal wastewater after passage through about 5 ft of soil. The coefficient 2 in the denominator is based on experimental data where 2 g of wastewater carbon were required to denitrify 1 g of wastewater nitrogen (11).

Nitrogen removal is also related to infiltration rate as shown in experiments with secondary effluent at Phoenix, Arizona (13). Lance showed that although nitrogen removal was 30% at an infiltration rate of 12 in./d, the removal increased to 80% at a

6 in/d infiltration rate. Based on this research, an application rate of 6 in/d is recommended as a maximum where 80% nitrogen removal is needed with secondary effluent. When primary effluent is used, the maximum application rate is recommended to not exceed 8 in/d. Because nitrogen removal is rarely required for RI, soil column testing or pilot testing with the actual wastewater and soil is recommended if these rates are to be exceeded.

To achieve the desired nitrogen removal, the application rate, or rate at which wastewater is discharged into the RI basins, may be less than the measured infiltration rate of the site. If this is the case, uniform applications with surface flooding of the basins may not be possible. In these cases, sprinkler distribution may be necessary.

Recent studies of nitrogen removal by RI lysimeters, applying secondary effluent, confirmed the work by Lance (14). At an application rate of 6 in/d, the total nitrogen removal was 80%. The optimum nitrogen removal was found with 1 day of flooding followed by 1 day of drying. In a full scale RI operation at Phoenix, the optimum removal of nitrogen was found with 9 days flooding and 12 days of drying (nearly a 1:1 ratio of flooding to drying).

In the same experiments with lysimeters, sprinkling for 15 min. followed by 75 min. of drying was not effective in nitrogen removal. Under these conditions only 16 to 23% of the nitrogen (mass basis) was removed, indicating that denitrifying conditions were not developed (14).

Phosphorus Removal

A conservative estimate of the phosphorus removal capability of an RI system can be made using Equation 3-3. The infiltration rate and flow distance determine the detention time. If the infiltration rate is too high to effect adequate phosphorus removal within an acceptable (site specific) flow path, the infiltration rate can be reduced by compacting the soil and by reducing the depth of wastewater applied. If the calculated phosphorus removal is not acceptable, a phosphorus adsorption test should be conducted. The results of the test should be multiplied by a factor of 5 to account for the slow precipitation that will occur over time. Phosphorus removal can also be tested using mathematical models (15)(16).

HYDRAULIC LOADING RATE

Selecting the appropriate design hydraulic loading rate is the most critical step in the process design procedure. As indicated in Chapter 7, an adequate number of measurements must be made of the infiltration rate and of the subsurface permeability. The hydraulic loading rate is a function of the site specific hydraulic characteristics, including infiltration, percolation, lateral flow, and depth to groundwater, as well as quality of the applied wastewater and the treatment requirements.

Design Infiltration Rate

The tests for infiltration rate described in Chapter 7 should be reviewed and an appropriate test selected. Using the equations in Chapter 4 (4-3 or 4-4) the mean infiltration rate is then calculated from the field data. During preliminary design the infiltration rate can be estimated from the SCS permeability data which is based on soil texture. For final design, however, actual field data should be used.

Wet/Dry Ratio

Intermittent application is critical to the successful operation of all land treatment systems. The ratio of wetting to drying in successful RI systems varies, but is always less than 1.0. Typical wet/dry ratios are presented in Table 12-2. For primary effluent the ratios are generally less than 0.2 to allow for adequate drying and scarification/removal of the applied solids. For secondary effluent, the ratio varies with the treatment objective, from 0.1 or less where nitrification or maximum hydraulic loading is the objective, to 0.5-1.0 where nitrogen removal is the treatment objective. These drying periods are necessary to restore the infiltration capacity and to renew the biological and chemical treatment capability of soil system.

Table 12.2: Typical Wet/Dry Ratios for RI Systems

Location	Preapplication Treatment	Application Period ....(days)	Drying Period	Wet/Dry Ratio
Barnstable, MA	Primary	1	7	0.14
Boulder, CO	Secondary	0.1	3	0.03
Calumet, MI	Untreated	2	14	0.14
Ft. Devens, MA	Primary	2	14	0.14
Hollister, CA	Primary	1	14	0.07
Lake George, NY	Secondary	0.4	5	0.08
Phoenix, AZ	Secondary	9	12	0.75
Vineland, NJ	Primary	2	10	0.20

Design Hydraulic Loading Rate

The design hydraulic loading rate for RI systems depends on the design infiltration rate and the treatment requirements. The procedure is to calculate the hydraulic loading rate based on a percentage of the test infiltration rate. This value is then compared to the loading rate based on treatment requirements and the lower rate is selected for design.

The most commonly used measurements for infiltration rates are the basin infiltration test and the cylinder infiltrometer (see Chapter 7). the relationship between annual loading rate and operating infiltration rates and cylinder infiltrometer rates is shown in Table 12-3.

Table 12.3: Typical Hydraulic Loading Rates for RI Systems (17)

Location	L (ft/yr)	.... Annual Loading Rate % of Operating Infiltration Rate	.... % of Cylinder Infiltration Rate
Boulder, CO	100-160	10-38	4-10
Brookings, SD	78-118	16-24	–
Ft. Devens, MA	95	13	2
Hollister, CA	50	24	3
Phoenix, AZ	200	27	–
Vineland, NJ	70	–	1.6

The saturated vertical hydraulic conductivity is a constant with time, whereas infiltration rates decrease as wastewater solids clog the soil surface. Thus, vertical conductivity measurements overestimate the wastewater infiltration rates that can be maintained over long periods of time. For this reason, and to allow adequate time for drying periods and for proper basin management, annual hydraulic loading rates should be limited to a fraction of the measured clear water permeability of the most restrictive soil layer.

Basin infiltration tests are the preferred method. However, their small area compared to the full scale basin, allows a larger fraction of the wastewater to flow horizontally through the soil from the test site than from the operating basin. Therefore, test infiltration rates are higher than the rates operating systems would achieve. Thus, design annual hydraulic loading rates should be no greater than 7 to 10% of measured basin test infiltration rates.

Cylinder infiltrometers greatly overestimate operating infiltration rates. When cylinder infiltrometer measurements are used, annual hydraulic loading rates should be no greater than 2 to 4% of the minimum measured infiltration rates. Annual hydraulic loading rates based on air entry permeameter test results should be in the same range.

Design guidance for hydraulic loading rates is summarized in Table 12-4. Where high wet/dry ratios and mild climates are expected, the upper end of the range of values in Table 12-4 can be used. Conversely, where long drying periods are needed, the lower end of the range should be used.

Table 12.4: Suggested Hydraulic Loading Rates Based on Different Field Measurements

Field Measurement	Annual Loading Rate
Basin infiltration test	7 to 10% of minimum measured infiltration rate
Cylinder infiltrometer and air entry permeameter measurements	2 to 4% of minimum measured infiltration rate
Vertical hydraulic conductivity measurements	4 to 10% of conductivity of most restricting soil layer

Example 12-1: Hydraulic Loading Rate

Conditions: The RI site consists of loamy sand. Basin infiltration tests yielded an average infiltration rate of 6.2 in/h. Determine the annual hydraulic loading rate.

Solution: Use an average of 8.5% to calculate the annual loading:

$$6.2 \text{ in/h} \times 24\frac{\text{h}}{\text{d}} \times 365\frac{\text{d}}{\text{yr}} \times \frac{1 \text{ ft}}{12 \text{ in.}} \times 0.085 = 385 \text{ ft/yr}$$

Before selecting this rate for design, check the treatment requirements and calculate the subsurface flow rate.

LAND REQUIREMENTS

The application area for RI systems can be determined using Equation 12-2.

$$A = \frac{Q(3.06)(365)}{L_w} \quad (12\text{-}2)$$

where
- A = application area, acres
- Q = average design flow, mgd
- 3.06 = conversion, acre-ft/mgd
- 365 = days/yr
- L_w = annual hydraulic loading, ft/yr

Other land requirements include area for preapplication treatment, roads, berms, and storage (if necessary). Access roads, typically 10 to 12 ft wide, are needed so that maintenance equipment for surface scarification can enter each basin. Storage is generally unnecessary for RI systems. The equivalent of short storage for emergencies can be attained by making the basins deep enough so that some storage can be realized. Area for future expansion should also be considered.

HYDRAULIC LOADING CYCLE

Loading cycles are selected to maximize either the infiltration rate, nitrogen removal, or nitrification. To maximize infiltration rates, the engineer should include drying periods that are long enough for soil reaeration and for drying and oxidation of filtered solids.

Loading cycles used to maximize nitrogen removal vary with the level of preapplication treatment and with the climate and season. In general, application periods must be long enough for soil bacteria to deplete soil oxygen, resulting in anaerobic conditions.

Nitrification requires short application periods followed by longer drying periods. Thus, hydraulic loading cycles used to achieve nitrification are essentially the same as the cycles used to maximize infiltration rates.

Recommended hydraulic loading cycles are summarized in Table 12-5. Generally the shorter drying periods in Table 12-5 should only be used in mild climates. In cold climates the longer drying periods should be used.

Table 12.5: Suggested RI Loading Cycles

Loading Cycle Objective	Applied Wastewater	Season	Application Period*	Drying Period
			.... (days)	....
Maximize infiltration rates	Primary	Summer	1-2	5-7
		Winter	1-2	7-12
	Secondary	Summer	1-3	4-5
		Winter	1-3	5-10
Maximize nitrogen removal	Primary	Summer	1-2	10-14
		Winter	1-2	12-16
	Secondary	Summer	7-9	10-15
		Winter	9-12	12-16
Maximize nitrification	Primary	Summer	1-2	5-7
		Winter	1-2	7-12
	Secondary	Summer	1-3	4-5
		Winter	1-3	5-10

*Regardless of season or cycle objective, application periods for primary effluent should be limited to 1 to 2 days to prevent excessive soil clogging.

Number of Basin Sets

The number of basins or sets of basins depends on the topography and the hydraulic loading cycle. The decision on the number of basins and the number to be flooded at one time affects both the distribution system hydraulics and the final wet/dry ratio. As a minimum, the system should have enough basins so that at least one basin can be flooded at all times (see Chapter 9). The minimum number of basins required for continuous wastewater application is presented in Table 12-6 as a function of the loading cycle.

Table 12.6: Minimum Number of Basins Required for Continuous Wastewater Application

Loading Application Period ....(days)	Cycle Drying Period ....(days)	Minimum Number of Infiltration Basins
1	5-7	6-8
2	5-7	4-5
1	7-12	8-13
2	7-12	5-7
1	4-5	5-6
2	4-5	3-4
3	4-5	3
1	5-10	6-11
2	5-10	4-6
3	5-10	3-5
1	10-14	11-15
2	10-14	6-8
1	12-16	13-17
2	12-16	7-9
7	10-15	3-4
8	10-15	3
9	10-15	3
7	12-16	3-4
8	12-16	3
9	12-16	3

Application Rate

The application rate is set by the annual loading rate and the loading cycle. The application rate is used to determine the required hydraulic capacity of the piping to the basins. The application rate is calculated as follows:

1. Add the application period to the drying period to obtain the total cycle time, days.
2. Divide the number of application days per year, usually 365 except where storage is planned, by the total cycle

time to obtain the number of cycles per year.

3. Divide the annual hydraulic loading by the number of cycles per year to obtain the loading per cycle.
4. Divide the loading per cycle by the application period to obtain the application rate, ft/d.

The discharge rate to the basins can then be determined using Equation 12-3.

$$Q = 18.9\ AR \qquad (12\text{-}3)$$

where Q = discharge capacity, gal/min

18.9 = conversion constant

A = basin area, acres

R = application rate, in/d

Example 12-2: Discharge Capacity

Conditions: The annual hydraulic loading rate for an RI system is 100 ft/yr. The application period is 1 day, the drying period is 13 days, and the basin area is 2 acres. Determine the application rate and discharge capacity.

Solution:

1. Total cycle time = 1 + 13 = 14d
2. Number of cycles per year = 365/14 = 26
3. Loading per cycle = 100/26 = 3.85 ft/cycle
4. Application rate = 3.85/1 = 3.85 ft/d
5. Q = 18.9 (2 acres)(3.85)(12 in/ft) = 1,746 gal/min.

COLD WEATHER OPERATION

In regions that experience cold weather, longer loading cycles may be necessary during winter months. Nitrification, denitrification, oxidation (of accumulated organics), and drying rates all decrease during cold weather, particularly as the temperature of the applied wastewater decreases. Longer application periods are needed for denitrification so that the application rate is reduced as the rate of nitrogen removal decreases. Similarly, longer resting periods are needed to compensate for reduced nitrification and drying rates.

Where ponds are used as preapplication treatment with cold winter weather, winter storage may be required. This is because the temperature of the wastewater becomes quite low prior to land treatment and makes the applied wastewater susceptible to long-term freezing in the basin. Alternatively, RI may be continued through cold weather if warmer wastewater from the first cell of the pond system (if possible) is applied.

Rapid infiltration systems that operate successfully during cold winter weather without any cold weather modifications can be found in Victor, MT; Calumet, MI: and Fort Devens, MA. However, some modifications have been used to improve cold weather treatment in other communities. Basin surfaces that are covered with grass or weeds should be mowed during fall. Mowing followed by disking should prevent ice from freezing to vegetation near the soil surface. Floating ice helps insulate the applied wastewater, whereas ice that freezes at the soil surface prevents infiltration. Problems with ice freezing to vegetation have been reported at Brookings, SD, where basins were not mowed and lagoons are used for preapplication treatment.

Another cold weather modification involves digging a ridge and furrow system in the basin surface. Following wastewater application, ice forms on the surface of the water and forms bridges between the ridges as the water level drops. Subsequent loadings are applied beneath the surface of the ice, which insulates the wastewater and the soil surface. For bridging to occur, a thick layer of ice must form before the wastewater surface drops below the top of the ridges. This modification has been used successfully in Boulder, CO and Westby, WI.

The third type of basin modification involves the use of snow fencing or other materials to keep a snow cover over the infiltration basins. The snow insulates both applied wastewater and soil.

DRAINAGE

Rapid infiltration systems require adequate drainage to maintain infiltration rates and treatment efficiencies. The infiltration rate may be limited by the horizontal hydraulic conductivity of the underlying aquifer. Also, if there is insufficient drainage, the soil will remain saturated with water and reaeration will be inadequate for oxidation of ammonia nitrogen to occur.

Renovated water may be isolated to protect either or both the groundwater or the renovated water. In both cases, there must be some method of engineered drainage to keep renovated water from mixing with native groundwater.

Natural drainage often involves subsurface flow to surface waters. If water rights are important, the engineer must determine whether the renovated water will drain to the correct watershed or whether wells or underdrains will be needed to convey the renovated water to the required surface water. <u>In all cases, the engineer needs to determine the direction of subsurface flow due to drainage from RI basins</u>.

Subsurface Drainage to Surface Waters

If natural subsurface drainage to surface water is planned, soil characteristics can be analyzed to determine if the renovated water will flow from the recharge site to the surface water. For subsurface discharge to a surface water to occur, the width of the infiltration area must be limited to values equal to or less than the width calculated in the following equation (18):

$$W = KDH/dL \qquad (12\text{-}4)$$

where W = total width of infiltration area in direction of groundwater flow, ft

K = permeability of aquifer in direction of groundwater flow, ft/d

D = average thickness of aquifer below the water table and perpendicular to the direction of flow, ft

H = elevation difference between the water level of the water course and the maximum allowable water table below the spreading area, ft

d = lateral flow distance from infiltration area to surface water, ft

L = annual hydraulic loading rate (expressed as daily rate), ft/d

Examples of these parameters are shown in Figure 12-1.

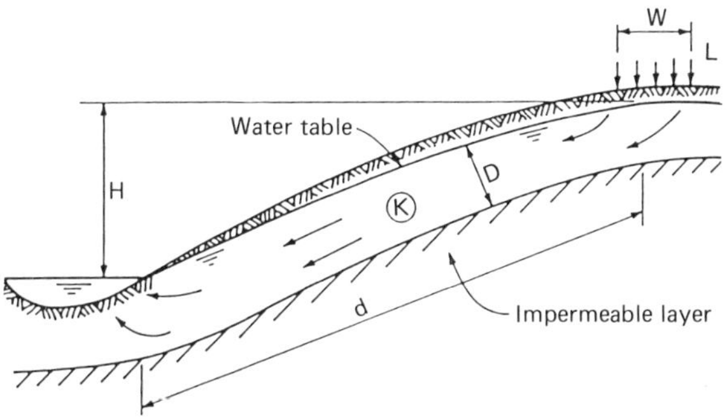

Figure 12-1 Definition sketch for lateral drainage from RI systems.

Example 12-3: Subsurface Drainage

Conditions: An RI site is located near a river with expected subsurface flow from the RI site to the river. The aquifer below the site is 20 ft thick and has a K = 3 ft/d. The annual hydraulic loading is 60 ft/yr. The water elevation is 30 ft below the RI basins and the lateral flow distance is 100 ft. If the groundwater mound is to be maintained at 5 ft or more from the RI basin surface, what is the maximum width of the RI basin area?

Solution: The maximum elevation difference, H, is 30-5 =

25 ft. The annual loading rate expressed as a daily rate is 60/365 = 0.16 ft/d.

$$W = \frac{KDH}{dL}$$

$$W = \frac{(3)(20)(25)}{(100)(0.16)}$$

$$W = 94 \text{ ft.}$$

Underdrains

Excessive groundwater mounding will inhibit infiltration and reduce the effectiveness of treatment. For this reason, the capillary fringe above the groundwater mound should never be closer than 2 ft to the bottom of the infiltration basin (19). This distance corresponds to a water table depth of about 3 to 7 ft, depending on the soil texture. The distance to groundwater should be 5 to 10 ft below the soil surface within 2 to 3 days following a wastewater application. Procedures for estimating groundwater mounding and underdrain spacings are provided in Chapter 4.

Generally, drains are spaced 50 ft or more apart and are at depths of 8 to 16 ft. In soils with high lateral permeability, spacing may approach 500 ft. Although closer drain spacing allows more control over the depth of the groundwater table, as drain spacing decreases the cost of providing underdrains increases. When designing a drainage system, different values of d should be selected and used to calculate S, so that the optimum combination of d, H, and S can be determined. Detailed information on drainage may be found in the U.S. Bureau of Reclamation Drainage Manual (20) and in the American Society of Agronomy manual, Drainage for Agriculture (21).

Once the drain spacing has been calculated, drain sizing should be determined. Usually, 6 in. or 8 in. drainage laterals are used. The laterals connect to a collector main that must be sized to convey the expected drainage flows. Drainage laterals should be placed so that they will be free flowing; the engineer should check drainage hydraulics to determine necessary drain slopes.

Recovery Wells

Rapid infiltration systems that utilize unconfined and relatively deep aquifers should use wells if necessary to improve drainage or to remove renovated water for reuse. Wells are used to collect renovated water directly beneath the RI sites at both Phoenix, AZ and Fresno, CA. Wells are also involved in the reuse of recharged waste water at Whittier Narrows, CA; however, the wells pump groundwater that happens to contain reclaimed water, rather than pumping specifically for renovated water.

The arrangement of wells and recharge areas varies; wells may be located midway between two recharge areas, may be placed on either side of a single recharge strip, or may surround a central infiltration area. Well design is described in detail in reference (22).

REFERENCES

1. Baillod, C.R., et al., Preliminary Evaluation of 88 Years of Rapid Infiltration of Raw Municipal Sewage at Calumet, Michigan, In: Land as a Waste Management Alternative, Ann Arbor Science, (1977).
2. Roy F. Weston, Inc., Operation and Maintenance Considerations for Land Treatment Systems, EPA-600/2-82-039, (January, 1982).
3. Satterwhite, M.B., Condike, B.J., and Stewart, G.L., Treatment of Primary Sewage Effluent by Rapid Infiltration, U.S. Army Corps of Engineers, Cold Regions Research and Engineering Laboratory (December 1976).
4. Pound, C.E., Crites, R.W., and Olson, J.V. Long-Term Effects of Land Application of Domestic Wastewater-Hollister, California, Rapid Infiltration Site. EPA-600/2-78-084, (April 1978).

5. Aulenbach, D.B., Long-Term Recharge of Trickling Filter Effluent into Sand. EPA-600/2-79-068, (March 1979).

6. Benham-Blair and Affiliates, Inc., Long-Term Effects of Land Application of Domestic Wastewater: Milton, Wisconsin, Rapid Infiltration Site, EPA-600/2-79-145, (August 1979).

7. Bouwer, H., et al., Rapid Infiltration Research at Flushing Meadows Project, Arizona, Journal WPCF, 52(10):2457-2470, (1980).

8. Pound, C.E. and Crites, R.W., Wastewater Treatment and Reuse by Land Application, EPA-660/2-73-006n, (August 1973).

9. Koerner, E.T. and Haws, D.A., Long-Term Effects of Land Application of Domestic Wastewater: Vineland, New Jersey, Rapid Infiltration Site, EPA-600/2-79-072, (March 1979).

10. Dryden, F.D. and Chen, C., Groundwater Recharge with Reclaimed Waters from the Pomona, San Jose Creek, and Whittier Narrows Plants, In: Proceedings of International Symposium on Land Treatment of Wastewater, Volume 1, U.S.A. CRREL, Hanover, NH, p. 241, (August 1978).

11. Leach, L.E., Enfield, C.G., and Harlin, C.C., Jr., Summary of Long-Term Rapid Infiltration System Studies, EPA-600/2-80-165, (July 1980).

12. Carlson, R.R., et al., Rapid Infiltration Treatment of Primary and Secondary Effluents, Journal WPCF, 54(3):270-280 (1982).

13. Lance, J.C., Whisler, F.D., and Rice, R. C., Maximizing Denitrification During Soil Filtration of Sewage Water, Journal of Environmental Quality, 5:102, (1976).

14. Leach, L.E. and Enfield, C.G., Nitrogen Control in Domestic Wastewater Rapid Infiltration Systems, Journal WPCF, 55 (9): 1150-1157, (1983).

15. Ryden, J.C., Syers, J.K., Iskandar, I.K., Evaluation of a Simple Model for Predicting Phosphorus Removal by Soils During Land Treatment of Wastewater, U.S. Army Corps of Engineers, CRREL, Special Report 82-14, (June 1982).

16. Enfield, C.G., Evaluation of Phosphorus Models for Prediction of Percolate Water Quality in Land Treatment, Proceedings of the International Symposium on Land Treatment of Wastewater, Volume 1, CRREL, Hanover, NH, p. 153, (August 1978).

17. U.S.EPA, Process Design Manual, Land Treatment of Municipal Wastewater, EPA 625/1-81-013, (October 1981).

18. Bouwer, H., Infiltration-Percolation Systems, In: Proceedings of the Symposium on Land Application of Wastewater, Newark, Delaware, p.85, (November 1974).

19. Bouwer, H., Zoning Aquifers for Tertiary Treatment of Wastewater, Ground Water, 14(6):386, (1976).

20. Drainage Manual, U.S. Department of Interior, Bureau of Reclamation, (1978).

21. Drainage for Agriculture, van Schifgaarde, J. ed, American Society of Agronomy, Series on Agronomy, No. 17, (1974).

22. Campbell, M.D. and Lehr, J.H., Water Well Technology, New York: McGraw-Hill, Inc., (1973).

13

Cost and Energy Considerations

COSTS

There are eight major categories of capital costs for land treatment systems:

1. Transmission
2. Pumping
3. Preapplication Treatment
4. Storage
5. Field Preparation
6. Distribution
7. Recovery
8. Land

In addition, there are costs associated with monitoring, administration buildings, roads and service and interest factors. There also may be costs for fencing, relocation of residents and purchase of water rights. Depending on the site management, SR and OF systems may have costs associated with crop planting, cultivating and harvesting.

There are operation and maintenance (O&M) costs associated with all of the eight categories of capital costs except for land and field preparation. These O&M costs can be divided into categories of labor, power and materials. Labor and materials for distribution and recovery are presented in this chapter. Power costs for pumping can be estimated from the energy requirements. All costs in this Chapter are for July 1984 and should only be used for approximating order-of-magnitude costs in planning.

Transmission

Transmission of wastewater to application sites can involve gravity pipe, open channels or force mains. Pumping can also be involved with gravity flow transmission but is required for force main transmission. Costs of transmission depend on the pipe or channel size and can be estimated using references 1 and 2.

Pumping

Pumping facilities for land treatment, as described in Chapter 9, range from full pumping stations to tailwater pumping facilities (see Recovery). Capital costs of transmission pumping depend on the type of structure. For example, a fully enclosed wet well/dry well structure, pumps, piping and valves, controls and electrical work will cost $300,000 for a 1 mgd peak flow and a 150 ft head (1). For structures built into the dike of ponds, the costs of pumping for distribution would be approximately $210,000 for the same flow and head conditions using an ENR Construction Cost Index (ENR CCI) of 4200.

Preapplication Treatment

Preapplication treatment for land treatment (Chapter 8) ranges from preliminary screening to advanced secondary treatment. Where a completely new land treatment system is to be constructed, it will usually be cost-effective to minimize preapplication treatment and use screening for OF systems and treatment ponds for SR and RI systems. Costs of preapplication treatment can be estimated from data in references (1)(2)(3). Many processes can be used for preapplication treatment including using OF as preapplication treatment for RI or SR systems. For example, at Shafter, CA an OF system was found to be cost-effective preapplication treatment for a municipal SR system (4).

Storage

Storage ponds vary in cost depending on initial site conditions, and the depth and volume of wastewater to be stored. Cost data is available in standard reference sources (1)(2)(3).

Field Preparation

Costs for field preparation can include site clearing and rough grading, land leveling and overland flow slope construction. Costs of each of these types of field preparation are presented in Table 13-1 for various conditions. Site clearing costs include bulldozing of existing vegetation, rough grading, and disposal of debris onsite. Offsite disposal of debris will cost 1.8 to 2.2 times the values in Table 13-1. Land leveling costs include surveying, earthmoving, finish grading, ripping two directions, disking, equipment mobilization, and landplaning. In many cases 200 yd^3/acre will be sufficient, while 750 yd^3/acre represents considerable earthmoving.

Overland flow slope construction costs include the same items as for land leveling. A cut of 500 yd^3/acre would correspond to nominal construction on pre-existing slopes. A cut of 1,000 yd^3/acre corresponds to constructing 150 ft wide slopes at 2 percent slope from initially level ground. A cut of 1,400 yd^3/acre corresponds to 250 ft slope widths on 2.5 percent slopes from initially level ground.

Table 13.1: Costs of Field Preparation

ENR CCI = 4,200

	Capital Cost ($/acre)
Site Clearing	
Grass only	20
Open field, some brush	150
Brush and trees	1,000
Heavily wooded	3,000
Land Leveling	
200 yd^3/acre	250
500 yd^3/acre	500
750 yd^3/acre	700
Overland Flow Slope Construction	
500 yd^3/acre	900
1,000 yd^3/acre	1,500
1,400 yd^3/acre	2,000

Distribution

Slow rate systems are capable of using a wide variety of sprinkler and surface distribution systems. In contrast, OF systems usually employ fixed sprinkler or gated pipe surface distribution and RI systems generally employ surface spreading basins.

Solid set sprinkling, described in Chapter 9, is the most expensive type of sprinkler system. As shown in Table 13-2, portable and continuous-move systems are considerably less expensive on an initial capital cost basis. Capital and O&M costs are presented in detail for solid set and center pivot sprinkling.

Table 13.2: Comparison of Sprinkler Distribution Capital Costs (5)

Sprinkler Type	Comparative Cost
Portable hand move	0.13
Traveling gun	0.22
Side roll	0.22
Center pivot	0.50
Linear move	0.65
Solid set	1.00

Solid Set Sprinkling: The capital and O&M costs for buried solid set systems are presented in Figure 13-1. For the SR system the laterals are spaced 100 ft apart and the sprinklers are 80 ft apart on the lateral. Laterals are buried 18 in. and mainlines are buried 36 in. The pipe material is PVC while the risers are galvanized steel. Flow to the laterals is controlled by hydraulically operated automatic valves. There are 5.4 sprinklers per acre at the specified spacing. If more sprinklers are included (smaller spacing), increase the capital and labor costs by using Equation 13-1:

$$\text{Cost Factor} = 0.68 + 0.06\,S \qquad (13\text{-}1)$$

where cost factor = multiplier times cost from Figure 13-1

S = sprinklers/acre

For overland flow the slopes are 250 ft wide and at a 2.5%

grade. The laterals are 70 ft from the top of the slope and sprinklers are 100 ft apart. Other factors are the same as for the SR system.

For O&M the labor rate, including fringes, is $10.00/h. Materials cost includes replacement of sprinklers and valve controllers every 10 yr.

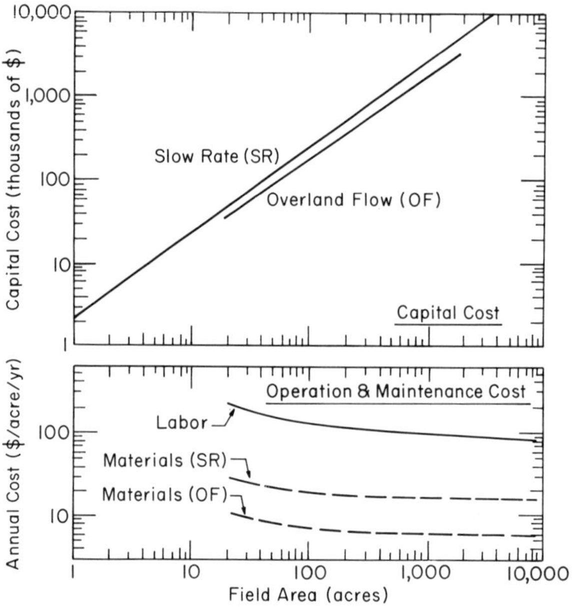

Figure 13-1 Solid set sprinkler (buried) costs, ENRCCI = 4200.

Center Pivot Sprinkling: Capital and O&M costs for center pivot sprinkling are presented in Figure 13-2.

The center pivot machines are electrically-driven and heavy-duty units. Multiple units are included for areas over 40 acres with a maximum area per unit of 132 acres. Distribution piping is buried 36 in.

Labor costs are based on $10.00/h and power costs are based

on 3.5 d/wk operation for each unit and $0.06/kwh. Materials cost includes minor repair parts and overhaul of units every 10 yr.

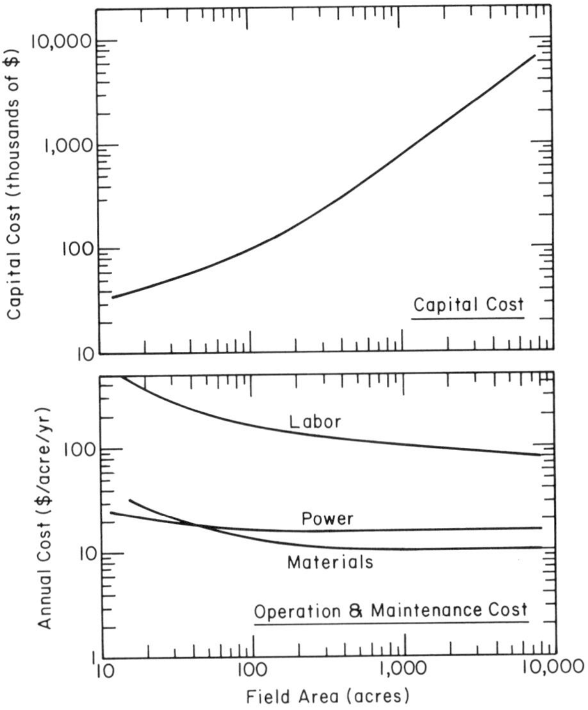

Figure 13-2 Center pivot sprinkling costs, ENR, CCI = 4200.

Surface Distribution for OF or SR: Costs for gated pipe distribution for OF and SR systems are presented in Figure 13-3. The OF slope is 200 ft wide with the gated aluminum pipe distribution at the top of the slope. For SR systems the furrows or borders are 1,200 ft long on rectangular-shaped fields. Graded border systems, under similar conditions of border length, can use buried pipelines with alfalfa valves (see Figure 9-6) at similar capital costs.

Labor costs are based on a $10.00/h wage including fringes. Materials cost includes replacement of gated pipe after 10 yr.

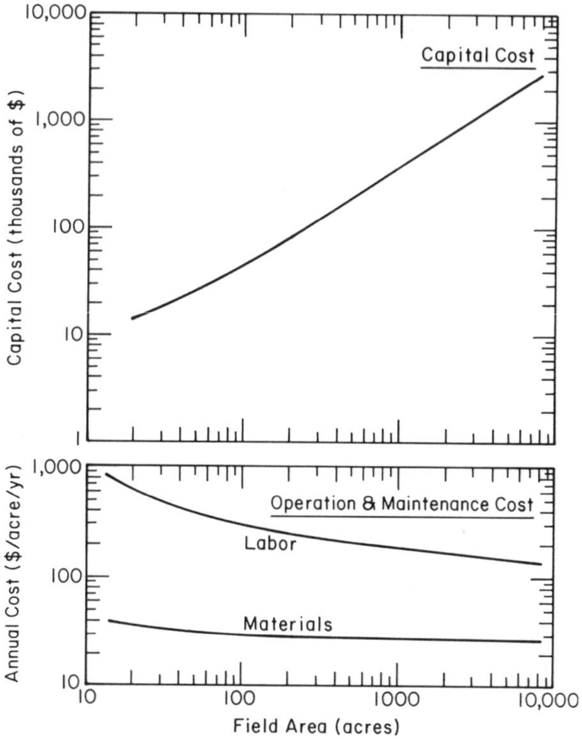

Figure 13-3 Gated pipe - OF or ridge and furrow SR, Costs, ENR CCI = 4200.

Rapid Infiltration Basins: Costs for RI basins are presented in Figure 13-4.

There are a minimum of 2 basins and a maximum size of 20 acres for an individual basin. Costs include inlet and outlet systems and control valves. Dikes are 4 ft high with an inside slope of 3:1, an outside slope of 2:1 and a 6 ft wide dike crest. Dikes or berms are formed from excavated native material.

Labor costs are based on a $10.00/h wage including fringe benefits. Materials cost includes rototilling or disking basin surfaces every 6 months and major repair of dikes every 10 yr.

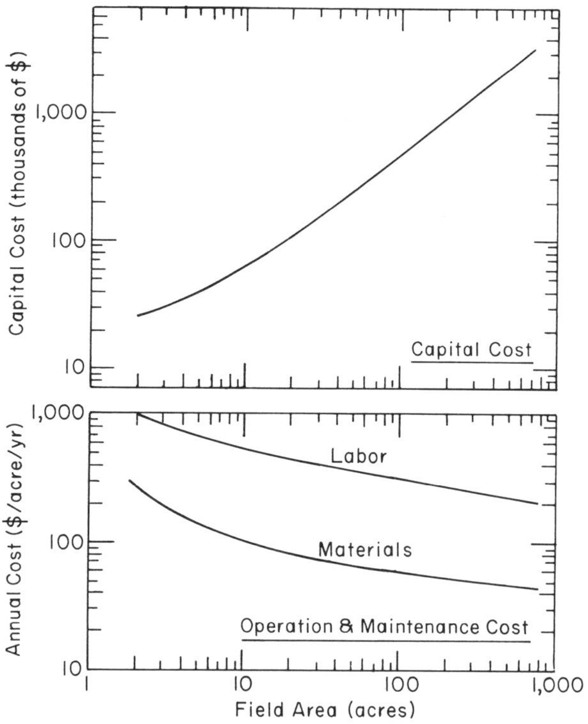

Figure 13-4 Rapid infiltration basin costs, ENR CCI = 4200.

Recovery

Recovery systems can include underdrains (for SR or RI), tailwater return for SR surface application, runoff collection for OF, and recovery wells for RI.

Underdrains: Costs for underdrain systems are presented in Table 13-3 for spacings between drains of 100 and 400 ft. Drains are buried 6 to 8 ft deep and discharge into an interception ditch along the length of the field.

Labor costs are based on a $10.00/h wage rate including fringe benefits and labor involves inspection and unclogging of drains at the outlets. Materials cost includes high pressure jet cleaning of drains every 5 yr, annual cleaning of interception ditch, and major repair of the ditch after 10 yr.

Table 13.3: Costs of Underdrains

ENR CCI = 4,200

Capital Costs	$/acre
100 ft spacing	2,000
400 ft spacing	750
O&M Costs	**$/acre·yr**
Labor	
100 ft spacing	35
400 ft spacing	15
Materials	
100 ft spacing	100
400 ft spacing	60

Tailwater Return: Tailwater from ridge and furrow or graded border systems must be recycled either to the storage ponds or to the distribution system. Typically 25% of the applied flow should be expected as tailwater. Costs, presented in Table 13-4, include drainage collection ditches, storage sump, pumping facilities, and 200 ft of return force main. Labor, at $10.00/h, includes operation of the pumping system and maintenance of the ditches, sump, pump, and return system. Materials cost includes major repair of the pumping station after 10 yr. Power cost is based on $0.06/kwh.

Table 13.4: Costs of Tailwater Return Systems

ENR CCI = 4,200

0.1 mgd of Recovered Water	Cost
Capital, $	40,000
O&M	
Power, $/yr	300
Labor, $/yr	250
Materials, $/yr	120
1.0 mgd of Recovered Water	
Capital, $	100,000
O&M	
Power, $/yr	3,600
Labor, $/yr	600
Materials, $/yr	500

Runoff Collection for OF: Runoff collection can consist of an open ditch or a buried pipeline with inlets. Costs for open ditches presented in Table 13-5, include a network of ditches sized for a 2 in/h storm, culverts under service roads, and concrete drop structures every 1,000 ft (for larger systems). For gravity pipe systems the costs include a network of interceptor pipes with inlets every 250 ft and manholes every 500 ft.

Labor costs are based on \$10.00/h including fringe benefits. Materials cost includes biannual cleaning of ditches and major repair very 10 yr.

Table 13.5: Costs of Runoff Collection for Overland Flow

ENR CCI = 4,200

Capital Costs	\$/acre
Gravity pipe system	1,600
Open ditch system	250
O&M Costs	**\$/acre·yr**
Labor	
Gravity pipe	5
Open ditch	20
Materials	
Gravity pipe	5
Open ditch	30

Recovery Wells: Costs for recovery of percolate with wells are presented in Table 13-6 for well depths of 50 and 100 ft. Capital costs include gravel-packed wells, vertical turbine pumps, simple shelters over each well, controls, and electrical work. Labor, at \$10.00/h, includes operation and preventative maintenance. Materials cost includes repair work performed by contract, and replacement of parts. Power cost is based on \$0.06/kwh.

Monitoring wells are generally a minimum of 4 in. in diameter. Wells typically cost \$30 to \$50/ft (1).

Table 13.6: Costs of Recovery Wells

ENR CCI = 4,200

1.0 mgd of Recovered Water	Cost
Capital, \$	
50 ft depth	20,000
100 ft depth	35,000
O/M	
Power, 50 ft depth, \$/yr	6,300
Power, 100 ft depth, \$/yr	12,600
Labor, \$/yr	4,000
Materials, \$/yr	600

Land

Land can be controlled by direct purchase, lease, or contract. The land for preapplication treatment and storage is usually purchased, however, field area for SR systems is sometimes leased or a contract is formed with the landowner. Options used by selected communities for land acquisition and management for SR systems are presented in Table 13-7.

Table 13.7: Options for Land Acquisition and Management at Selected SR Systems (6)(7)

Location	Site (acres)	Acquisition Option	Management Option
Bakersfield, CA	2,400	Fee simple	Leaseback to farmer
Camarillo, CA	475	Contract	Landowner accepts water
Dickinson, ND	250	Contract	Cash lease for water sale to farmer
Lubbock, TX	4,000	Fee simple and contract	Leaseback, farmer owns effluent
Mesa, AZ	160	Fee simple	Leaseback for cash rent
Muskegon, MI	10,400	Fee simple	Managed by county
Petaluma, CA	550	Contract	Cash rent for irrigation equipment
Roswell, NM	285	Contract	Cash lease for water sale to farmer
San Angelo, TX	740	Fee simple	Managed by city
Tooele, UT	1,200	Contract	Cash lease for water sale to farmer

As shown in Table 13-7 contracts for effluent use are used in several SR systems. Fee simple purchase is generally used for OF and RI sites.

Land Application of Sludge

The principal costs involved in land application of liquid or dewatered sludge are for hauling and applying the sludge. Truck hauling is the most popular method of sludge transport. Cost factors for trucking sludge are presented in Table 13-8 (8).

Table 13.8: Capital and Operating Costs of Sludge Hauling (8)

ENR CCI = 4,200

Sludge Type	Truck Type	Capacity	Capital Costs ($ x 1,000)	Operation Costs ($/mi)
Liquid	Tank truck	1,200 gal	45-50	0.38
		2,500 gal	75-85	0.50
		5,500 gal	115-130	0.58
Dewatered	Dump truck	8-10 yd^3	45-50	0.38
		10-15 yd^3	85-90	0.50
		15-25 yd^3	100-110	0.58
	Bottom dump truck	25 yd^3	125-140	0.70

BENEFITS

Revenue producing benefits from land treatment systems can include sale of crops, lease of land, sale of wastewater or reclaimed, and contracts that may involve all of these benefits. Examples of revenue producing benefits are presented in Table 13-9. The examples are for SR systems, which generally have the greatest potential for revenue production. Crop sale from OF systems can offset a small portion of O&M costs, but generally cannot be expected to more than offset the cost of harvesting. For RI systems in water-short areas the potential for recovery and reuse of the percolate should be investigated.

Table 13.9: Benefits of Land Treatment Systems (5)(6)(7)(9)(10)

Sale of Crops	\$/yr
Muskegon, MI	900,000-1,000,000
San Angelo, TX	58,000-71,000
Lease of Land	**\$/acre·yr**
Bakersfield, CA	80
Coleman, TX	5
Manteca, CA	40
Mesa, AZ	50
Winters, TX	20
Sale of Effluent	**\$/acre·ft**
Cerritos, CA	40
Irvine Ranch, CA	118
Las Virgenes, CA	160
Marin MWD, CA	300

Sale of crops can be a significant source of revenue if the community is willing to invest in the necessary equipment for crop harvest. For example, Muskegon County has realized gross revenues of \$1,000,000 in 1976 and \$900,000 in 1977 and 1978 from the sale of corn grain (9).

Cash rent for SR cropland is very popular in the West with 5 year agreements being common. Rents range from \$5 to \$80/acre.

Contracts for wastewater use, use of irrigation equipment, or for grazing of cattle on pasture are also widely used. Examples of contracts include El Reno, OK, Dickinson, ND, Mitchell, SD, Tuolumne County, CA, Santa Rosa, CA and Petaluma, CA (7)(11).

ENERGY REQUIREMENTS

Energy requirements for land treatment systems include power for pumping, preapplication treatment, and wastewater distribution and fuel for crop planting and harvesting and for sludge transport and spreading. In addition energy is needed for heating and cooling of buildings, lighting and vehicle operation.

Pumping

Pumping for transmission, distribution, tailwater return, and recovery is a major energy use in most land treatment systems. The energy required can be calculated using Equation 13-2:

$$\text{Energy, k/Wh/yr} = \frac{(Q)(TH)(t)}{(F)(E)} \qquad (13\text{-}2)$$

where
- Q = flowrate, gpm
- TH = total head, ft
- t = pumping time, h/yr
- F = constant, 3960 x 0.746 = 2954
- E = overall pumping efficiency, decimal

The overall efficiency depends on the type of wastewater and the specifics of pump and motor selection. In the absence of specific information on pump and motor efficiency, the following overall pumping system efficiencies can be used:

Raw wastewater	0.4
Primary effluent	0.65
Secondary effluent, tailwater, recovered groundwater	0.75

Land Treatment of Wastewater

Distribution energy can be calculated using Equation 13-2.

Energy for preapplication can be estimated from references (12) (13). Energy for crop production is minor compared to energy for distribution. For example, energy requirements for corn production are 5.7 kwh/acre and for alfalfa are 2.5 kwh/acre. Fuel usage can be converted to energy using 124,000 Btu/gal for gasoline and 14,000 Btu/gal for diesel (3)(14).

Land Application of Sludge

Transport and spreading of sludge requires fuel energy. For example, if 5 M gal/yr of liquid sludge is hauled 20 mi (one-way) in a 2,500 gal capacity truck, a total of 80,000 mi/yr would be driven. If the truck gets 4.5 mi/gal of diesel, the 17,800 gal/yr of fuel would be equivalent to 2.5 x 10^9 Btu/yr.

ENERGY CONSERVATION

Sprinkler distribution systems are candidates for energy conservation. Impact sprinklers may require 150 to 200 ft of head to operate. Recent advances have been made in sprinkler nozzle design to allow operation at lower pressures without sacrificing uniformity of application. Use of drop nozzles with pressure requirements of 50 ft of head can result in significant energy conservation.

Energy conservation is also possible in land treatment systems through the use of surface distribution. A comparison of primary and secondary energy usage of various land and aquatic treatment systems is presented in Table 13-10.

Table 13.10: Energy Requirements for Land and Aquatic Treatment Systems (3)

	Equivalent Energy, 10^3 kWh/yr		
System	Primary Energy	Secondary Energy	Total Energy
PT + RI	187	102	289
Ponds and wetlands	121	198	319
PT + SR (surface)	187	135	322
PT + OF	192	159	351
Ponds and hyacinths	167	195	362
PT + SR (spray)	327	173	500

Note: PT = primary treatment, RI = rapid infiltration, SR = slow rate and OF = overland flow.

Energy conservation through the use of land application of wastes can also be realized through savings in energy use for manufacturing of commercial fertilizer. A presentation of energy needs to produce fertilizer and the energy value of nutrients in wastewater is given in Table 13-11 (14).

Table 13.11: Energy Value of Nutrients in Wastewater (14)

Nutrient	Content of . . . Effluent . . . mg/ℓ	lb/acre·ft	Energy to Produce, Transport and Apply Fertilizer (kWh/lb)	Energy Value of Nutrients in Wastewater (kWh/acre·ft)
Nitrogen as N	20	54	2.79	190
Phosphorus as P	10	27	0.10	13
Potassium as K	15	38	0.10	10

REFERENCES

1. Reed, S.C., et al., Cost of Land Treatment Systems, EPA 430/9-75-003, (September 1979).
2. U.S.EPA, Construction Costs for Municipal Wastewater Conveyance Systems: 1973-1979, EPA 430/9-81-003, (January 1981).
3. Tchobanoglous, G., Colt, J.E., and Crites, R.W., Energy and Resource Consumption in Land and Aquatic Treatment Systems, Proceedings, Energy Optimization of Water and Wastewater Management for Municipal and Industrial Applications Conference, USDOE, Volume 2, New Orleans, (December 10-13, 1979).
4. Corneille, R.W. and Pound, C.E., Overland Flow Treatment: A Case History at Shafter, CA, Proceedings, National Seminar on Overland Flow Technology for Municipal Wastewater, Dallas, EPA-600/9-81-022, (September 1980).
5. Crites, R.W., Economics of Reuse, Proceedings, Water Reuse Symposium II, Volume 3, AWWA, p. 1745-1751, (August 23-28, 1981).
6. Christensen, L.A., Irrigating With Municipal Effluent, USDA, ERS-672, Washington, D.C., (January 1982).
7. Crites, R.W., Innovative and Alternative Treatment at Petaluma, CA, Presented at the Hawaii Water Pollution Control Association Annual Conference, Honolulu, HI, (January 28, 1982).

8. U.S.EPA, Process Design Manual Land Application of Municipal Sludge, EPA-625/1-83-016, (October 1983).
9. Walker, J.M., Wastewater: Is Muskegon County's Solution Your Solution?, EPA-905/2-76-004, (August 1979).
10. Roy F. Weston, Inc., Operation and Maintenance Considerations for Land Treatment Systems, EPA-600/2-82-039, (January 1982).
11. National Association of Conservation Districts, The Role of Conservation Districts and the Agricultural Community in Wastewater Land Treatment, EPA-430/9-81-019, (September 1981).
12. Wesner, G.M., et al., Energy Conservation in Municipal Wastewater Treatment, EPA-430/9-77-011, (1978).
13. Middlebrooks, E.J. and Middlebrooks, C.H., Energy Requirements for Small Flow Wastewater Treatment Systems, USA CRREL Special Report 79-7, (April 1979).
14. Water Pollution Control Federation, Energy Conservation in the Design and Operation of Wastewater Treatment Facilities, Manual of Practice No. FD-2, (1981).

14

Operation, Maintenance and Monitoring

The proper operation and maintenance (O&M) of land treatment systems is essential for the realization of performance expectations. Land treatment systems are not more labor intensive than conventional wastewater technologies. However, a broader range of skills may be required for those land treatment systems that incorporate an agricultural or silvicultural component.

The mechanical elements (ie: pumps, valves, etc.) that are common to all wastewater systems are not discussed in this chapter. The major focus is on the unique aspects of O&M and monitoring for land treatment systems.

SLOW RATE SYSTEMS

The type of SR system can range from a remote forested site with no public access to a golf course or park with frequent public use. Agricultural systems can be managed by the industry or the municipality, or the wastewater can be delivered to private farmers for their use. Each of these systems will have different requirements for O&M. Both the type of system and the management plan are determined during design but there is the potential for subsequent change. For example, a change from forest to crop production or a decision to allow public access on the site may then require higher levels of pretreatment and disinfection.

Staffing Requirements

The number of operating personnel and the skill levels required will depend on the type of system and on its size. Figure 14-1 presents an estimate of the personnel needs of typical municipal SR land treatment systems. The figure shows the approximate number of hours per day for the smaller systems and the number of full-time employees required for the larger systems. These estimates are only for the land treatment portion of the system. These estimates are for a "typical" system; an agricultural operation producing row crops will require more time for O&M than indicated, a forested site will require less.

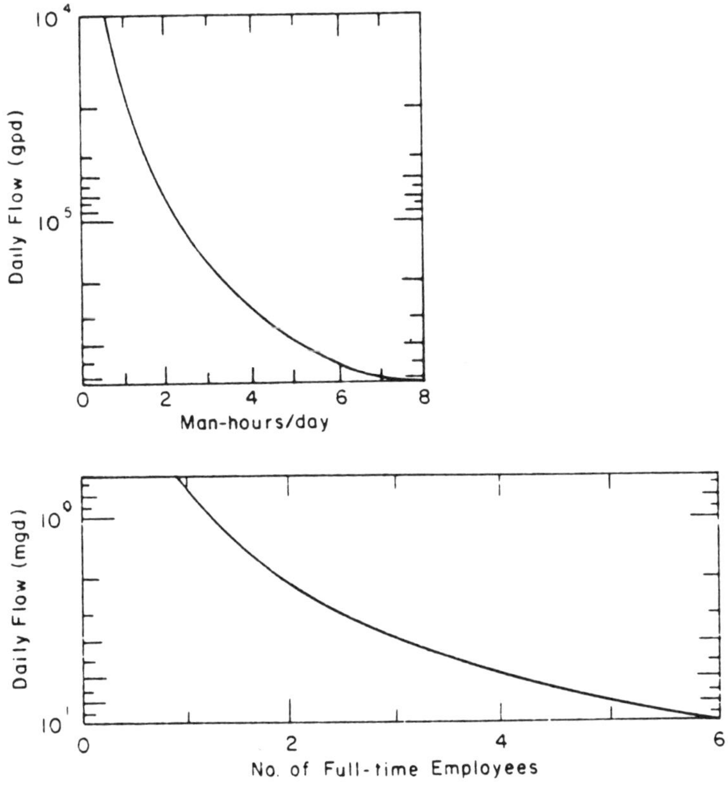

Figure 14-1 Personnel needs for land treatment portion of SR systems, After (5).

General Skills

The general skills required for routine operation of all types of SR systems are essentially the same as those needed for routine operation of any simple waste treatment system. A unique requirement for SR land treatment is deciding when to turn the water on and off or when to switch the application to a different part of the site. A basic program and schedule of operations will have been determined for each project during final design. However, this may require adjustment by the operator if flow increases, if a year is especially wet or dry, or if the vegetation used in the system is changed.

Special Skills

The operator of a forested site will sometimes need expert advice to help with problems such as insect infestations or plant diseases, or to determine which trees to cull or when to clearcut. The operator of an agricultural site will require all of the farming skills normally associated with the particular type of crop (pastures, hay crops, or row crops).

Recreational sites require particular attention to water quality to maintain adequate health protection. In addition, the wastewater application scheduling for recreational sites requires careful control so as not to interfere with recreational activities. That will usually involve nighttime or off-season application. Many recreational sites will include a carefully maintained turfgrass cover.

Process Control and Monitoring

The information needed for operation of the system is obtained through the monitoring program. Monitoring needs can be divided into two categories. There is compliance monitoring to certify that the system is meeting the requirements of the Federal, state and local agencies that are responsible. There is also routine process monitoring to ensure that all internal components in the system are functioning as designed. This type of monitoring is necessary if regulatory requirements do not exist. However, it is often possible to satisfy both regulatory and operating needs at the same time if the monitoring program is planned carefully.

Compliance Monitoring: The Federal government and all states have regulations controlling discharges to surface waters. Land treatment systems that collect the treated water with underdrains or wells and then discharge it to surface waters will need a permit, as will most overland flow systems. Although a discharge permit may not be required for the case where the treated water remains in the ground or emerges into surface water at some remote place, these systems are not ignored by the regulatory agencies. Many states now require permits to discharge to groundwater. Their criteria range from very specific regulations that have the force of law, to general guidelines that may be strongly recommended but which are more flexible in application than regulations. There are also case-by-case determinations that depend on the site conditions and operational plan of a particular system.

The U.S. EPA guidelines for the level of preapplication treatment believed suitable for various types of land treatment systems are given in Table 8-1. All of the 50 states have an interest in and some level of control over the monitoring of land treatment systems, even if there is no surface discharge. The monitoring requirements for a particular system will have been determined during design and will be written into the O&M manual.

Monitoring Requirements: The majority of states are concerned about the quality of the wastewater to be applied, and many have specific regulations or guidelines. Groundwater protection is a case-by-case concern and it depends largely on the groundwater use in the vicinity of the facility and the classification of the aquifer. Monitoring of the soil is usually of the process control type to make sure the system operates properly or to warn of long-term effects that might inhibit the future use of the site for other purposes. Crops may also be monitored for operational purposes. The potential for aerosol contamination is of little concern to most state agencies, except on a case-by-case basis for recreational operations and those that are close to the public.

A typical example of the type and frequency of monitoring required for the applied wastewater is shown in Table 14-1. The BOD, pH, nitrogen and phosphorus are familiar water quality parameters tested in most systems. Many other parameters (metals, etc.) are not shown in Table 14-1. Either they are not generally present in sufficient concentration in the typical domestic-municipal wastewaters or their presence has no direct effect on the proper operation of the system.

Table 14.1: Typical Monitoring Schedule for Applied Wastewater

Parameter	...Size of System (mgd)..	
	0-1.0	>1.0
BOD	Q	M
Suspended solids (SS)	Q	M
pH	Q	W
Kjeldahl-nitrogen	Q	W
Ammonium-nitrogen	Q	M
Nitrate-nitrogen	A	M
Phosphorus	A	Q
Potassium	A	Q
Sodium	A	Q
Calcium	A	Q
Magnesium	A	Q
Chloride	A	M
Total dissolved solids	A	M

Note: Q = quarterly, A = annually, M = monthly, and W = weekly.

Groundwater Monitoring: Groundwater is usually monitored at the system boundaries when the quality of drinking water aquifers is a factor. Nitrate-nitrogen (NO_3^--N) is the parameter of greatest concern, but it is advisable to measure the organic and ammonia nitrogen as well since they can subsequently be oxidized to nitrate. In general, SR systems that can remove enough nitrogen to meet drinking water standards at the project boundaries will also remove all of the other constituents of concern in typical municipal wastewaters. Frequent sampling is not necessary because groundwater moves relatively slowly and rapid changes in quality will not be observed. Samples taken once or twice per year should be sufficient. The design of those systems that operate only seasonally should include an estimate of the travel time for the percolate to reach the project boundary and the sampling operation should be scheduled accordingly.

Since there may be little vertical mixing of the groundwater and the system percolate, the sampling depth of the monitoring wells must be carefully selected during design. Wells that are too deep will probably not obtain samples that have been influenced by the land treatment operation. The location and depth of monitoring wells should be determined during design. However, it may be necessary to add new wells if operational conditions change or if groundwater levels were not properly determined during design. Figure 7-4 illustrates the design features for a typical shallow monitoring well that, depending on soil conditions, could be installed to depths of 10-15 ft by the system operator. Deeper wells will generally require mechanical drilling techniques.

Groundwater monitoring wells are sometimes installed within the application site as well as at the project boundaries. These wells monitor performance immediately beneath the application site and measure the depth to groundwater under the application site. Figure 14-2 illustrates one relatively easy technique for measuring the depth to water in monitoring wells. Since samples are taken infrequently from these monitoring wells, the water standing in the casing will not be representative of the true groundwater quality. At least three casing volumes should be pumped, or removed with a well bailer, prior to well sampling.

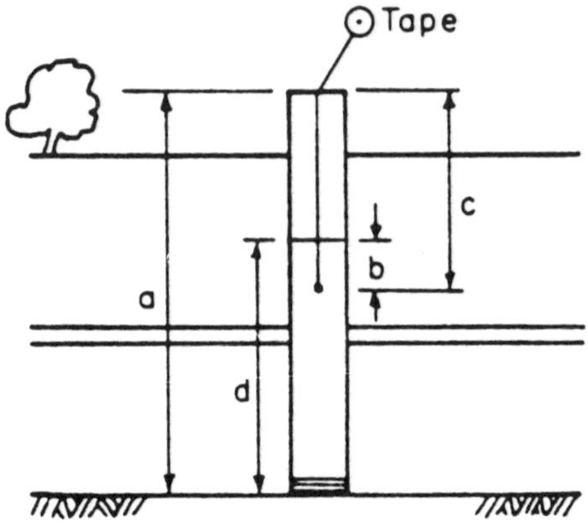

a = top of casing elevation above datum
b = length of wetted tape
c = tape reading--read exactly at the top of the casing
d = piezometric head, relative to a given datum
c-b = dtw (depth to water)
a-dtw = d (piezometric head at the center of the screen, relative to indicated datum plane).

Figure 14-2 Water level determination in observation wells, After (2).

The location of monitoring wells is based on the determination of the groundwater flow direction made during system design. As shown in Figure 14-3 the perimeter wells are installed on the hydraulic downgradient of the site. In addition a monitoring well should be installed on the upgradient side to measure water quality before the groundwater flows beneath the site.

Measuring the groundwater elevation in these wells can confirm that the direction of flow is as predicted in design.

Springs or seeps in unexpected locations after the system starts up are usually a sign of groundwater movement, and additional wells may be needed in those directions.

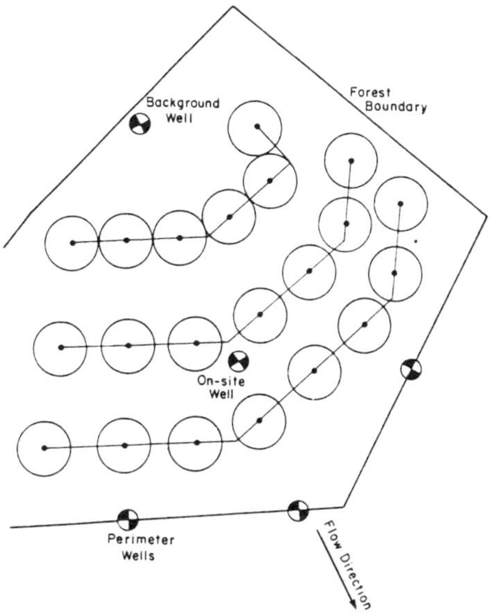

Figure 14-3 Typical monitoring well layout, After (5).

Storage Ponds

Many of the newer SR systems combine preapplication treatment and storage in a single pond system. Monitoring needs include regular measurement of water level in the storage pond as well as water quality tests just before and during the period of land application.

Water level in the pond should be measured at least weekly during the operating season. The method of observation can range from a simple marker board or staff gauge visually observed, to automatic, and sometimes transmitting, water level recorders. Direct observation by the operator is recommended, even if automated equipment is installed, to allow him to also observe dikes and other pond structures.

In any particular year there may be more or less water in the storage pond than was predicted during design either due to changes in wastewater flows or to extremes in precipitation. The operator must then revise the application schedule accordingly to make certain that the vegetation on the site gets enough water and also to achieve the specified pond water level at the end of the season. Usually, the pumping system has been designed to deliver a certain flow and is not adjustable. However, the operator can vary the operating time for the pumps, start the application season earlier, extend it, or change the amount of water put on particular parts of the site. Suggestions for appropriate action on each case are listed below:

Operation Procedures for More Water Than Normal in Storage:

1. Forest, pasture and hay crop sites. Start application earlier (as soon as frost is out of the ground) and extend the season into late fall. If different soils exist, apply more water to areas with coarser soils by increasing pumping time. Apply more water to the entire site by increasing pumping time, but do not allow ponding or runoff of wastewater.
2. Agricultural row crops. Continue application for longer period after crop harvest. Consult with county agricultural extension agent and plant a more water tolerant crop that year. Increase application to the maximum amount recommended by the extension agency for the crop grown. Plant a rye grass mixture on the coarsest soils on the site. Continue normal row cropping and application practice on rest of the site. Apply at highest possible rates on the grassed plot, and plow under grass and return to normal practice the following year.
3. Recreational sites. Increase the application period to the maximum possible without interfering with public access, and/or restrict access to a portion of the site and apply at higher rates on that por-

tion, and continue application after the recreational season has ended.

Operation Procedures for Less Water Than Normal in Storage:

1. Forest, pasture and hay crop sites. In arid climates, reduce the amount to be applied per week but continue applications to the whole site. In humid climates, take a portion of the site out of service, continue the application on the rest at design rates; if vegetation on the out-of-service portion shows stress then apply some water.
2. Agricultural row crops. In both arid and humid climates, calculate how much water is available for application, and determine the water needs per acre of the crop to be grown. Plant only the number acres that can be supported with available flow.
3. Recreational sites. Reduce the amount to be applied per week but continue applications to the whole site. In dry climates this will probably require extra water the following year to leach salts from the root zone if the application has been reduced severely.

Application Site Monitoring

Monitoring at the application site is necessary to ensure that the system operates properly. Monitoring tasks will include observing the sprinklers, pumps and other mechanical equipment and determining soil fertility and crop quality at agricultural sites. These requirements usually apply to all sites and essentially consist of routine visual observations and record-keeping. With seasonally operated systems it is essential to know when the soils thaw in the spring and when they freeze in the winter if these factors control the application schedules developed during design. The actual time of freezing and thawing will vary from year to year and may be different than the design assumptions. An especially heavy rainfall during the application season may require adjustment in the routine weekly schedule. If the amount of rainfall from a single storm or closely spaced storms is equal to the amount of wastewater scheduled for application it may be necessary to delay applications for a few days so that there is no runoff. The operator must also observe areas where there might be ponding in low spots. These shallow puddles can lead to odor and insect problems and must be eliminated. Watching the sprinkler patterns will reveal clogged nozzles or other mechanical problems in the system. If the site is underdrained the drain outlets should be routinely inspected to make certain that they are flowing. If monitoring wells exist on the site, the depth of water in the well should be regularly measured. In general, if the groundwater table gets within 5 ft of the surface, wastewater applications should be temporarily reduced or stopped.

Longer-term monitoring is to insure soil fertility and good crop quality, and this requires periodic sampling and testing. Table 14-2 presents a suggested monitoring program for the soils at an agricultural site. The number of samples taken will depend on the number of different soil types at the site and the size of the site. Specific guidance can be obtained from the county agricultural extension agent, but for the general case there should be a composite soil sample representing each of the major soil types.

Baseline samples should be taken and tested either during the final stages of design or just before the system is put into operation. A pH determination is needed to see if lime is required to adjust soil pH for the crop to be grown (See Figure 7-16). Phosphorus and potassium results are needed to decide if supplemental fertilization is required. These tests should be repeated annually for high value crops; however, once every 3 years is suitable for hay and similar crops. The county agricultural extension agent can help interpret these test results and tell the operator how to correct any problems.

Table 14.2: Soil Monitoring on Agricultural Sites (5)

Annual Sample and Test	Baseline and Every 5 Years
pH (for lime needs)	pH (for lime needs)
	Nitrogen
Available phosphorus	
Exchangeable/extractable	Cation exchange capacity, % organic matter
Potassium	
Sodium	Exchangeable/extractable
Magnesium	Potassium
Calcium	Phosphorus
	Copper
	Zinc
	Nickel
	Cadmium
	Total
	Boron
	Copper
	Zinc
	Nickel
	Cadmium
	Sodium adsorption ratio (SAR) in western states or arid climates

Table 14-3 lists the suggested parameters and frequency of tests for vegetation monitoring at agricultural sites. If a system is designed for nitrogen or phosphorus removal, then the total nitrogen and phosphorus concentration in the harvested crops should be measured and crop yield should be determined. This will allow calculation of removal performance by the crop to ensure that the system is functioning as designed. If forage grasses or silage are the crop and if these are fed to livestock, then high nitrate content in grasses may cause health problems in the livestock. The analysis may be important if wastewater with a high nitrogen content (>30 mg/L) is used and if the application season has been unusually wet and cool. The county agricultural extension agent should be consulted for advice on the testing need in a particular year. The vegetation should be tested for the metals listed in Table 14-3 at the same frequency as the soil tests to establish long-term trends.

Table 14.3: Vegetation Monitoring on Agricultural Sites (5)

Component	Frequency
Total nitrogen and phosphorus	Annual sample if N or P removal is required by system
Nitrate (NO_3) for forage grasses and silage	At harvest if recommended by extension agent for livestock protection
Copper, zinc, nickel, cadmium	At first harvest and every 5 years thereafter to establish trends

The number of samples required for these tests and the part of the plant to sample are critical to the reliability of results. Hay cuttings and green chop for silage from small fields can be sampled immediately after harvest since a representative composite sample can be obtained from the mixed materials. Large or scattered fields and other crops should be sampled in the field with recommended sampling patterns. It is best to sample the leaves rather than the plant fruits since the leafy matter will usually show increase in metal content first and thereby given an earlier warning of potential problems. Table 14-4 recommends techniques for vegetation sampling.

Table 14.4: Vegetation Sampling–Field Pattern and Plant Part (5)

Crop	Pattern	Plant Part
Alfalfa	X diagonals of field, 50-100 clumps	Upper stem cutting in early flower stage
Corn	X diagonals or along row at least 50 plants into field	Center one-third of leaf, just below lower center; at full tassle
Wheat and grains	X diagonals, 200 or more leaves	First four leaf blades from top of plant
Grass and sod	X diagonals	Clippings or whole tops
Soybeans	Random leaves from at least 5% of plants, 50 to 100 leaves	Youngest mature leaves, after pod formation
Tree fruits	X diagonals in orchard, one leaf from north, south, east and west sides of tree	Mature leaves, shoulder height, 8 to 12 weeks after full bloom

Routine Operating Procedures: Factors of concern include application rates and schedules, crop management and the unique requirements for forested and recreational sites.

Application rates and schedules: Control of the water to be applied is common to all systems and requires the following operator decisions:

- Startup and shutdown schedule
- Quantity of wastewater to be applied each shift
- Frequency of application
- Field or section to be used

The details of these decisions may change from year to year depending on the climate, rainfall and type of crop but the final result must be to apply the total amount of wastewater required during the application season. A specific program will have been formulated during design and instructions will be included in the O&M manual. However, the operator must have the knowledge and the capability to alter the schedule to accommodate special conditions.

Year-round Operations: At sites where the wastewater is treated year-round and where there is usually little storage volume, the operator has little flexibility to make adjustments. The daily land treatment applications must match the daily wastewater flow. In most cases the operator can decide which field to use and how long to continue the application to that field. During the start-up phase the operator should use the schedule provided by the design engineer. However, the design is often based on average site conditions. The operator should carefully watch each area to see if the water is rapidly infiltrating, ponding or running off. Some parts of the site may be able to take more water than the design value and some parts less. The operator can then make adjustments to put more water on the better soils and less on the poorer locations.

Seasonal Flow Operations: Systems where the wastewater flow is seasonal are not uncommon. These might include camp grounds, ski resorts and seasonal industries. Wintertime flow in cold climates will usually require wastewater storage for land application in the warm months. The operator needs to know the dimensions of the pond, the length of the application season, the amount of wastewater that will flow into the pond during the season, and an estimate of rainfall or evaporation during the season. With this information he can calculate the applications and time schedules for each week.

Year-round Flow, Seasonal Applications: The basic procedures are similar to those in the previous case except that the entire annual wastewater flow (stored flow plus daily wastewater generated) must be applied during the application season. In addition, if the site is designed for agricultural row crops, startup will usually come after planting and application will be stopped for harvest and cultivation. There must not be any erosion, but with many systems it is possible to resume application to the bare fields after harvest is complete and to continue until freezeup.

Crop Management: Management of the crop is a major requirement at agricultural sites. A particular crop is usually selected during design and planted early in the first year of operation. It may be possible, thereafter, to change the crop to either improve the performance of the system or to increase the value of the harvest. See Chapter 5 for detailed discussion on crop types and responses.

Cutting Management: The type of cutting management for harvesting forage grasses will depend on the desired level of nitrogen removal. If maximum yields and high nitrogen removal are desired, grasses should be cut more frequently and at the proper times: the initial cutting should be at the early heading stage of growth and subsequent cuttings should be every 4 to 5 weeks for the remainder of the season. The early heading stage will vary with climate but it will usually be sometime during the middle to late spring.

1. If lower nitrogen removals, in the range of about

160 lb/acre of nitrogen, are needed, then fewer cuttings are needed and operations costs can be saved. Initial cuttings for this purpose should be at the late flowering stage of growth, with one extra cutting toward the end of the growing season. With this cutting method, the majority of the nitrogen will be removed at the initial harvest.

2. Grasses should be managed properly so that they can survive at the site as long as possible. Under proper management, grasses at the site can persist for 3 years or more. They are usually invaded by weedy grasses, some of which are desirable. The weed quack-grass has performed well in SR systems in terms of nitrogen removal and forage quality. When undesirable weeds predominate, fields must be renovated to maintain treatment efficiency. When reseeding, use standard methods for field renovation along with desirable types of forage grasses.
3. With corn much of the nitrogen is removed during a short 4 to 6 week period in summer. This is usually between the knee-high and the tasseling stages of growth.
4. Growing another crop with the corn can improve nitrogen removal by an additional 40 to 80 lb/acre and can lower the percolate nitrogen concentration. In intercropping (see Chapter 5), corn is grown during the summer months, while a cereal crop (e.g. rye) or a forage grass (e.g. reed canarygrass) is grown during the spring and fall. The cereal or grass removes nitrogen during the slower corn uptake periods and thereby lengthens the application season. The disadvantages of this system are that the actual corn yields will be lower due to increased competition with

the grasses and that a higher level of management is required.

5. Turfgrasses, grown for sod production or maintained in a lawn, can remove nitrogen during the entire growing season. When started from seed for sod production, nitrogen removal will be lower. The sod is usually harvested after 12 to 18 months. Weekly mowings during periods of active growth are desirable.

Operations at Forested Sites: Operations at forested sites will require the same decisions regarding how much water is needed, where to apply it, and how long to let it run, that were discussed above. In general, the frost-free season is longer for forests than for an agricultural field, and in northern climates forest soils that have an early snow cover may not freeze at all. In these cases wastewater application can continue all winter. Winter operation requires quick drainage of exposed pipes at the end of the application period. If not planned for during design, the operator will have to install drains at all low spots in the piping system before attempting winter operation. Other operational requirements for forested sites relate to tree management and will require expert advice. Every 3 to 4 years, an experienced forester should tour the site and make recommendations on culling or harvest and other management practices that will ensure a healthy stand of trees.

Recreational Sites: Recreational systems have the same basic requirements as the cases previously discussed. They are more difficult to operate since the recreational function and schedule usually take precedence over the wastewater renovation. The operator has to plan his operational schedule for wastewater applications so as not to interfere with the recreational activities.

Emergency Procedures: A major concern is disruption of the operating schedule for wastewater applications because there is usually a limited amount of storage capacity available. Since emergencies cannot be predicted, it is prudent for the operator to keep some part of the available storage free. This may require pumping slightly more water than the average schedule would require during the early part of the season.

Extended power failures also disrupt operations. The design should have provided the capability for standby power at the pumping stations. Systems that use center pivot distribution rigs with electrical drive motors should also have a portable standby generator for direct field connection when required.

If treatment and storage ponds are part of the system and there is public access to the site or if the site is close to a community, odors may be a concern from time to time. Odors are not a problem from properly designed and operated land treatment systems, but they may be possible if wastewater characteristics or pond conditions change suddenly. The operator must be prepared to cope immediately with such a situation.

Application to the pond of a chemical such as potassium nitrate will supress the odors and allow time for the cause of the problem to be identified and corrected. A recommended procedure is to apply 100 lb/acre of potassium nitrate on the first day and then 50 lb/acre pond surface on each day thereafter if odors persist. The chemical should be applied in the wake of a motor boat.

Odors will not occur on the actual land application site unless septic wastes are used or if stagnant puddles and ponds of wastewater are allowed to stand. The latter will also be the cause of insect problems. The operator must routinely inspect the application site and eliminate these low spots by filling with new soil.

Maintenance Procedures

The dikes and berms for ponds will require regular maintenance. Earthern dikes must be checked regularly for muskrats and other burrowing animals. Soil-cement, plastic membrane or asphalt liners must be regularly inspected and repaired. Damage from waves or ice in the winter are the most common problems.

Systems that use sprinklers must have a regular schedule for inspection and cleaning. All lines and pipes in seasonal operations should be regularly drained, even if freezing is not expected, to avoid corrosion. In addition to sprinkler maintenance, the large center pivot rigs require attention to their tires and gear boxes. Tires should be checked for proper pressure and the gear boxes for proper lubrication at the start of the operation season.

OVERLAND FLOW

The runoff water quality from an overland flow slope will depend on the amount of time wastewater is allowed to remain in contact with the soil surface. The operator can control detention time on an existing slope by adjusting the application rate. Chapter 11 contains guidance on start-up of OF systems. Staffing requirements for an overland flow system can be determined with Figure 14-1.

Crop Management

The grass should be cut two or three times a year and removed from the slopes if nitrogen control is a design requirement. Before harvesting, each slope must be allowed to dry out so that equipment can travel over the soil surface without leaving ruts. Ruts could develop into channeling, especially if they are oriented downslope. The drying time necessary before mowing is usually about 1 to 2 weeks; however, this can vary depending on the soil and climatic conditions. After mowing, the hay should be dried before raking and baling. This may take another week or so depending on the weather.

Suggested monitoring programs for soils and vegetation are the same for OF as for SR systems. If the grass is used as fodder, samples may be required during each harvest, and may be analyzed for various nutritive parameters such as protein, fiber, total digestible nutrients, phosphorus and dry matter. These analyses can be conducted by the agricultural department of most state universities.

RAPID INFILTRATION

The general O&M requirements for RI systems are similar to those used at any earthen basin. The special requirement for RI is maintenance of the design infiltration capacity.

In order to minimize any problems with the basins the operator should inspect them daily, and record in the daily log sheets the depth of standing water in the various basins and the amount of time it takes them to drain. This will allow calculation of the wastewater infiltration rate and identification of those basins where the infiltration rate has decreased to a level where restoration of the surface is needed. The operator should inspect the berms of the infiltration basins frequently. Vegetation such as tree seedlings and brush should be removed. The operator should also note any signs of erosion on the berms, and inspect the hydraulic system used to apply the wastewater to the basins to determine if it is functioning properly. Low spots where wastewater can remain ponded should be filled in. During winter operations the entire system must be inspected, paying particular attention to problems of freezing and ice formation.

Restoring the basins to an acceptable infiltration capacity is normally accomplished by discing or scarifying the dry soil surface to break up the organic mat that develops. Another method is to completely remove the top layer of soil and replace it with a suitable soil. This method uses more labor and equipment than discing and it will also require large earth moving equipment. Care must be taken to limit the amount of vehicular traffic on the beds to reduce the amount of compaction of the soil layers.

In colder climates the operator should disc the dry surface

of the basins about once each year during the late summer and fall. This should keep the basins from clogging during the winter season. Chapter 12 contains additional guidance on winter operations.

SLUDGE SYSTEMS

The land application of municipal and industrial sludges have many O&M requirements that are similar to SR systems. Chapter 16 discusses in detail the design of the major sludge application concepts. The sludge systems differ from SR in the application methods and scheduling, more frequent soils monitoring for the LDP and a greater concern for nuisance issues such as odors and spillage.

Monitoring Requirements

Table 14-5 lists typical monitoring requirements for agricultural utilization of sludge at "agronomic" rates (see Chapter 16). The major parameters of concern are (1) pH maintenance at 6.5 to reduce potential metal migration, and (2) soil P and K if optimum crop yields are a project goal. Nitrate in groundwater is generally only a problem when the sludge application(s) exceed the N needs of the crop.

Table 14.5: Typical Site Monitoring Requirements for Sludge Applications at or Below Agronomic Rates* (6)

Soil pH	Soil Test for P and K**	NO_3^- in Groundwater	Cd in Crop
Yes (2)***	Yes (2)	No	No

*Numbers in parentheses refer to frequency of analysis: 2 = every 2 years.
**Soil test for available N can be used, if appropriate.
***Frequency depends on amount of N applied, depth to groundwater, and amount of leachate. Regulatory agencies will dictate frequency.

Application Scheduling on Agricultural Sites

The timing of sludge applications must correspond to farming operations, and is influenced by crop, climate, and soil properties. Sludge cannot be applied during periods of inclement weather. In some states, sludge cannot be applied to soils that are frozen or covered with snow. Soil moisture is a major consideration which impacts the timing of sludge application. Traffic on wet soils during or immediately following heavy rainfalls may result in compaction and reduced crop yields; muddy soils also make vehicle operation difficult. Application to frozen or snow-covered ground with greater than 3 percent slope may result in excessive runoff into adjacent streams. In addition, sludge applications must be scheduled around the tillage, planting, and harvesting operations for the crops grown.

Sludge Use on Disturbed Land

As described in Chapter 16 this concept involves the use of sludge as an organic soil amendment to restore disturbed land such as strip mined areas. The typical approach is to apply a single large sludge application, with the amount determined by the LDP which is often total allowable cadmium. The operations during the brief application period generally involves large trucks and earth moving equipment for spreading and incorporating the sludge.

Prior to sludge application, the surface should be roughened or loosened to offset the compaction caused during the site leveling or grading operation. This will help to improve the surface water infiltration and permeability, and slow the movement of any surface runoff and erosion. A heavy mining disk or chisel plow is typically necessary to roughen the surface. It is advisable that this be done along the contour.

The timing of sludge application depends on the climate, soil conditions, and growing season. It is generally not advisable to apply sludge to frozen or snow-covered ground, since it cannot be immediately incorporated and seeded. The sludge should not be applied during periods of heavy rainfall since this greatly increases the chances of surface runoff. Sludge should not be applied in periods of prolonged extreme heat or dry conditions, since considerable amounts of nitrogen will be lost before the vegetation has a chance to establish itself.

Sludge applications should be scheduled to accommodate the growing season of the selected plant species. If the soil conditions are too wet when sludge is applied, the soil structure may be damaged, bulk density increased, and infiltration decreased due to heavy vehicle traffic on the wet soil. This may increase the possibility of soil erosion and surface runoff.

If the area to receive sludge is covered under federal or state mining regulations, the sludge application must be scheduled to comply with the revegetation regulations. For example, in Pennsylvania mined land can be seeded in the spring as soon as the ground is workable, usually early in March, but seeding must terminate by May 15. Late summer seeding season is from August 1 until September 15, and sludge application and seeding of mined land covered by these regulations must comply with these requirements.

Application Methods

The best suited technique depends on the type of sludge (liquid or semi-solid) and on whether it is a surface or subsurface application. Table 14-6 provides guidance for surface applied liquid sludge, Table 14-7 for subsurface applied liquid sludge and Table 14-8 for dewatered sludges.

Table 14.6: Surface Application Method and Equipment for Liquid Sludges (6)

Method	Characteristics	Topographical and Seasonal Limitations
Tank truck	Capacity 500 to more than 2,000 gallons; it is desirable to have flotation tires; can be used with temporary irrigation set-up; with pump discharge can achieve a uniform application rate.	Tillable land; not usable at all times with row crops or on very wet ground.
Farm tank wagon	Capacity 500 to 3,000 gallons; it is desirable for wagons to have flotation tires; can be used with temporary irrigation set-up; with pump discharge can achieve a uniform application rate.	Tillable land; not usable at all times with row crops or on very wet ground.

Table 14.7: Subsurface Application Methods, Characteristics, and Limitations for Liquid Sludges (6)

Method	Characteristics	Topographic and Seasonal Limitations
Flexible irrigation hose with plow or disc cover	Use with pipeline or tank truck with pressure discharge; hose connected to manifold discharge on plow or disc.	Tillable land; not usable on very wet or frozen ground.
Tank truck with plow or disc cover	500 gal commercial equipment available; sludge discharge in furrow ahead of plow or disc mounted on rear on 4-wheel-drive truck.	Tillable land; not usable on very wet or frozen ground.
Farm tank wagon with plow or disc cover	Sludge discharged into furrow ahead of plow mounted on tank trailer; application of 170 to 225 wet tons/ac; or sludge spread in narrow band on ground surface and immediately plowed under; application of 50 to 120 wet tons/acre.	Tillable land; not usable on very wet or frozen ground.
Subsurface injection	Sludge discharge into channel opened by a chisel tool mounted on tank truck or tool bar; application rate 25 to 50 wet tons/ac; vehicles should not traverse injected area for several days.	Tillable land; not usable on very wet or frozen ground.

Table 14.8: Methods and Equipment for Application of Dewatered Sludges

Method	Characteristics
Spreading	Truck-mounted or tractor-powered box spreader (commercially available); sludge spread evenly on ground; application rate controlled by over-the-ground speed; can be incorporated by disking or plowing.
Piles	Normally hauled by dump truck; spreading and leveling by bulldozer or grader needed to give uniform application.

REFERENCES

1. Weston, Inc., Roy F., Evaluation of Operation and Maintenace Practices and Design Considerations of Land Application Systems, EPA 600/2-82-039, U.S. EPA, MERL, Cincinnati, OH, (1982).
2. U.S. EPA, Process Design Manual for Land Application of Wastewater, EPA 625/1-81-013, U.S. EPA, CERI, Cincinnati, OH, (1981).
3. U.S. EPA, Operations Manual-Stabilization Ponds, EPA 430/9-77-012, U.S. EPA, OWPO, Washington, DC, (1977).

4. U. S. EPA, Field Manual-Performance, Evaluating and Troubleshooting of Municipal Wastewater Treatment Facilities, EPA 430/9-78-001, U.S. EPA, OWPO, Washington, DC, (1978).

5. U.S.ACE, Engineer Manual-Land Treatment Systems Operation and Maintenance, EM1110-2-504, U.S. Army Corps of Engineers, Washington, DC (1983).

6. U.S. EPA, Process Design Manual-Land Application of Municipal Sludge, EPA 625/1-83-016, U.S. EPA, CERI, Cincinnati, OH, (1983).

15

Small Scale Systems and Innovative Concepts

The procedures in this chapter are intended for industries and communities with wastewater flows of 250,000 gpd or less and when the LDP is a routine parameter such as BOD, N, or hydraulic capacity. The basic objectives for any land treatment system are the same regardless of size, however the design of small systems should include special emphasis on the ease of operation and on minimizing construction and operating costs. Most communities in this size range cannot hire full-time treatment plant operators, and the treatment system must be capable of providing consistent, reliable treatment in the absence of frequent attention. In general, most treatment systems that meet these objectives are nonmechanical and have no discharge to surface waters.

The concepts discussed in this chapter include:

- Large scale septic tank-inground disposal systems.
- Small scale applications of the basic land treatment concepts (SR, OF, RI).
- Wetlands and other innovative use of the soil ecosystem for waste treatment.

The procedures for planning and design of small systems are similar to, but less detailed than, the requirements for large facilities as described in Chapters 6 to 12. Maximum use is made of local expertise and existing published information. The area Soil Conservation Service (SCS) staff, the county agent, and local farmers can all provide assistance and advice. In effect the procedures described in this chapter for SR, OF and RI systems reduce the cost and complexity of site investigation, planning and system design by increasing the safety factors involved. This approach is typically acceptable for most small systems. However, if land costs are high, or the LDP for design is not a routine parameter, then the detailed approaches described in earlier chapters should be used.

ONSITE SEPTIC TANK SYSTEMS

This is the most common form of onsite wastewater disposal in the United States and design procedures are well established for the typical single family unit. However, the concept is finding increasing use in commercial applications, public buildings, and cluster-type residential developments. In these cases the flow can easily exceed 1,000 gpd and often approaches 10,000 to 20,000 gpd for a single system.

The land component in these systems is typically considered a disposal operation so the LDP for design is the hydraulic capacity of the natural soils. Treatment does occur but since the wastewater application point is typically below the ground surface the responses involve the soil and not the surface vegetation. Chapter 3 can be used to estimate these subsurface treatment responses for the parameters of concern.

A few shallow auger borings and some variation of the familiar percolation test are usually the source of design information for a single family system. The severe limitations of the USPHS percolation test are widely recognized but it is still the most commonly used test for onsite systems, regardless of size. It is not an appropriate basis for the design of systems when the daily flow approaches or exceeds 1,000 gpd.

Hydraulic failure will occur when the system cannot accept and then transmit, via subsurface flow, the design wastewater volume. This can be caused by biological clogging at the application point interface, by high groundwater conditions, or by soils with unsuitable permeability in the horizontal direction. The latter two can be evaluated by a proper site investigation and the larger the system the more extensive should be the field investigation. Test pits are a key element in identifying soil conditions and groundwater locations and the methods described in Chapter 7 should be followed.

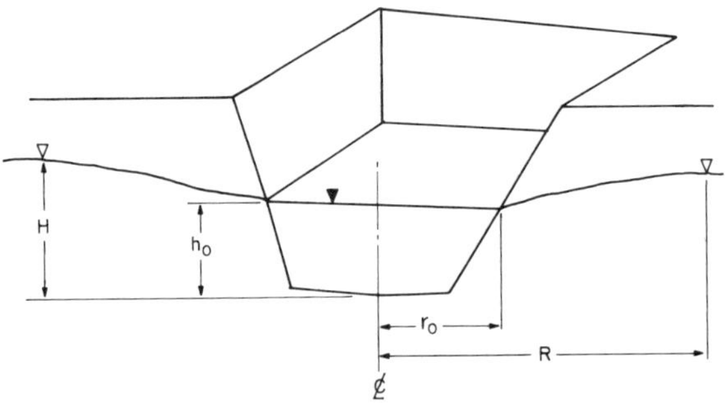

Figure 15-1 Definition sketch - pit permeability test.

Measurement or accurate estimates of the vertical and horizontal permeability of the various soil layers are necessary for design. Again, the procedures described in Chapter 7 can be applied. Healy and Laak (1) have developed a variation of the pump out auger hole test, using a test pit. Figure 15-1 illustrates the geometry of their test. The water table must be within 8 to 10 ft of the surface so it can be reached with a backhoe excavation. The test can either observe the rate of water level rise in the pit and a final determination of the equilibrium level or by waiting until the level stabilizes and then pumping out at least 1 ft of water and then observing the rate of rise. Since flow into the pit has both vertical and horizontal components, the permeability value determined is an overall "average" for the affected soil. This permeability can be determined with the following equations:

$$Q = \frac{\pi k(H^2 - h_o^2)}{2.3 \log (\frac{R}{r_o})} \qquad (15\text{-}1)$$

where Q = volumetric rate of flow, ft^3/h

$= \frac{\Delta h}{\Delta t} (A)$

$\frac{\Delta h}{\Delta t}$ = rate of water level rise, ft/h

A = area of water surface, ft^2 (may be different for each observation since hole is irregular)

k = "average" permeability of soil, ft/h

H = height to stable groundwater, ft

h_o = height to water level in pit at time t, ft

R = distance from pit centerline to stable water level, ft

r_o = distance from pit centerline to edge of water level in pit, ft

For the typical case $\frac{R}{r_o}$ can be assumed to be equal to 4. Substitution in Equation 15-1, and rearranging terms produces:

$$K = (\frac{\Delta h}{\Delta t}) \left[\frac{A}{2.27 (H^2 - h_o^2)}\right] \qquad (15\text{-}2)$$

Design Loading

Healy and Laak (1) suggest a design approach based on the long-term acceptance rate. They suggest that at a particular optimum hydraulic loading rate the accumulation of clogging biological materials will be in equilibrium with the decomposition rate so that flow could continue indefinitely under these conditions. Table 15-1 contains selected values from their work relating the soil permeability to the "long-term acceptance rate" in terms of feet of wastewater per year for domestic septic tank effluents, assuming a one foot head of water in the disposal trenches or bed. Most on site designs are based on a hydraulic loading expressed in terms of gallons per square foot per day. The relationship is given by:

$$L_{wg} = L_{wf}\ (0.0205) \qquad (15\text{-}3)$$

where L_{wg} = long term acceptance rate, $gal/ft^2 \cdot d$

L_{wf} = long term acceptance rate, ft/yr

Table 15.1: Soil Permeability–Disposal Field Wastewater Loading

k Soil Permeability (inch/hour)	Lf Acceptance Rate (feet/year)
0.14	13
0.30	15
0.70	18
1.4	19
2.9	21
7.0	27
14.0	38
30.0	63
70.0	120

A comparison of the values in Table 15-1, with Figure 4-6 in Chapter 4 indicates that the tabulated values are within the range shown on Figure 4-6. However, the coarse textured soils ($K > 2$ in/h) have much more conservative values in Table 15-1. This means that the biological clogging layer controls flow in the coarse soils with a subsurface point of wastewater application. The higher values permitted in Figure 4-6 are based on a surface application which allows for aerobic decomposition of the accummulated organics.

Designs for finer textured soil ($K < 2$ in/h) can use either Table 15-1 or Figure 4-6 to determine hydraulic loading; designs for coarser soil ($K > 2$ in/h) should be the values in Table 15-1. An alternative is to design for a higher value and then plan for periodic chemical restoration with hydrogen peroxide as described in a later section of this chapter. This approach may be necessary if the available area will not be sufficient for a large disposal field.

Example 15-1: Determine Application Bed Area for a 1,000 gpd Flow

Conditions: Test pit permeability results: $A = 10\ ft^2$, $H = 3.0\ ft$, $h_o = 2.5\ ft$, $\frac{\Delta h}{\Delta t} = 0.2\ in/h$.

Solution:

1. Use Equation 15-2.

$$K = \left(\frac{\Delta h}{\Delta t}\right) \frac{A}{2.27(H^2 - h_o{}^2)}$$

$$= (0.2)\ \frac{10}{2.27\ (9 - 6.25)}$$

$$= 0.32\ in/h$$

2. From Table 15-1.

with, $K = 0.32\ in/h$

use $L_{wf} = 15.5\ ft/yr$

3. Use Equation 15-3.

$$L_{wg} = (15.5)(0.0205) = 0.32\ gal/ft^2 \cdot d$$

$$\text{Bed area for 1,000 gpd} = \frac{1,000}{0.32} = 3,147\ ft^2$$

Groundwater Mounding

An estimate of the groundwater mounding that will occur beneath a large scale disposal bed or trench is necessary to insure successful performance. The detailed procedures in Chapter 4 for calculating mound characteristics can be used for onsite disposal systems as well as RI basins. A number of simplified calculation techniques have been developed for application to the smaller scale onsite diposal systems (2)(3)(4)(5). Only one of these methods is presented here as a demonstration. It is recommended that more than one procedure be used for very large scale systems to develop a range of possible conditions for final evaluation by the designer. Reference number (5) is particularly valuable in that respect since it presents several different models.

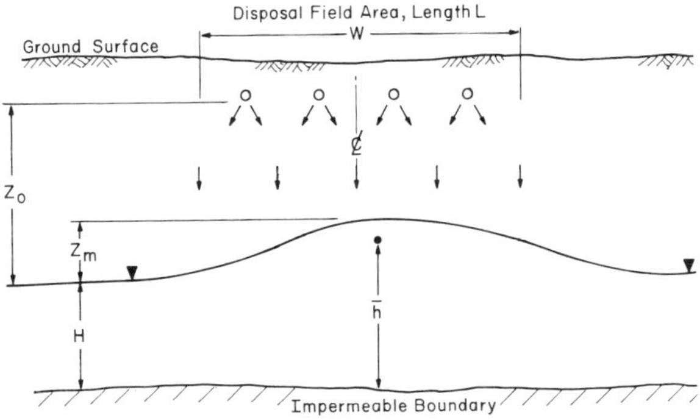

Figure 15-2 Definition sketch - mound geometry for simplified calculation procedure.

Figure 15-2 defines the mound geometry for the simplified calculation procedure developed by Finnemore and Hantzsche (2) based on theoretical considerations presented by Hantush (6). The related equations are:

$$\bar{h} = H + \frac{Z_m}{2} \qquad (15\text{-}4)$$

where $\bar{h}$ = distance from boundary to mid point of the long-term mound

H = height of stable groundwater table above impermeable boundary, ft

Z_m = long-term, maximum rise of the mound, ft

$$Z_m = \frac{QC}{A}\left[\frac{L}{4}\right]^n \left[\frac{1}{K\bar{h}}\right]^{0.5n} \left[\frac{t}{S_y}\right]^{1-0.5n} \qquad (15\text{-}5)$$

where: Q = average flow, ft^3/d

A = area of disposal field, ft^2

C = constant, see Table 15-2

L = length of disposal field, ft

K = horizontal permeability of soil, ft/d

$\bar{h}$ = see Equation (15-4)

n = exponent, see Table 15-2

S_y = specific yield of aquifer, see Chapter 4

t = time since beginning of wastewater application, d

The K value can be determined in the field with slug tests or bailing tests as described in Chapter 7. In the general case the horizontal conductivity of most soils is significantly higher than the vertical conductivity. Therefore, a conservative estimate of mound rise will be produced if the vertical conductivity is used in Equation 15-5. At very short time periods Equation 15-5 will predict high, but conservative, predictions of mound rise. A period of 10 years is recommended by the authors (2) for calculation purposes.

An iterative approach may be needed for the solution since it is necessary to assume a Z_m for Equation (15-4) to determine $\bar{h}$ so that Equation (15-5) can be solved for Z_m.

Table 15.2: Constants for Equation 15-5

Length to Width Ratio (L/W) of Disposal Field	C	n
1	3.4179	1.7193
2	2.0748	1.7552
4	1.1348	1.7716
8	0.5922	1.7793

Example 15-2: Determine Mound Height After 10 Years

Conditions: Wastewater discharge = 2,500 gpd, Disposal bed: L = W = 100 ft, H = 50 ft, S_y = 0.2, Soil permeability = 1 in/h, Z_o = 6 ft.

Solution:

$$\frac{L}{W} = \frac{100}{100} = 1, \text{ from Table 15-2: } C = 3.4179$$

$$n = 1.7193$$

$$K = (1\ \frac{in.}{h})\ (24\ \frac{h}{d})\ (\frac{1\ ft}{12\ in.}) = 2\ ft/d$$

$$Q = (2{,}500)(\frac{1}{2.48}) = 334\ ft^3/d$$

$$t = (10\ \text{yr})(365\ \frac{d}{yr}) = 3{,}650\ d$$

1. First estimate:

Assume $Z_m = 5.0$ ft

So: $\bar{h} = 50 + \frac{5}{2} = 52.5$ ft

Then using Equation (15-5):

$$Z_m = \frac{(334)(3.4179)}{10{,}000}\left[\frac{100}{4}\right]^{1.7193}\left[\frac{1}{(2)(52.5)}\right]^{0.8597}\left[\frac{3650}{0.2}\right]^{0.1404}$$

$$= (0.114)(253.2)(0.01876)(3.965)$$

$$= 2.14\ \text{ft}$$

2. Second iteration:

$$\bar{h} = 50 + \frac{2.14}{2} = 51.07$$

$$Z_m = (0.114)(253.2)(.01874)(3.965)$$

$$= 2.14\ \text{ft}$$

It is clear that the calculation for this example is not sensitive to the initial estimate of Z_m since H is very large compared to Z_m. In these cases it is acceptable to assume:

$$\bar{h} = H \tag{15-6}$$

The calculation is very sensitive when the groundwater depth (H) is relatively small. In the example, had H been only 12 ft then the mound would rise 6 ft in the 10 years for the conditions specified and the system would be in a failure mode since the original depth to groundwater (Z_o) was only 6 ft. The H value selected for use in Equation (15-5) should be the mean normal saturated depth of the aquifer. The mound height that is then determined should be added to the highest seasonal water table elevation for the site to obtain the worst case condition.

Changing the configuration of the disposal field may reduce the groundwater mounding. For example, in the previous case a square area 100 x 100 ft was assumed. Adopting a rectangular area of 50 x 200 ft would change the constants derived from Table

15-2 and, for the initial conditions specified in Example 15-1, the mound would rise 1.9 ft as compared to 2.1 ft for the square configuration.

Soil Mound Systems

Mound systems were developed to overcome site constraints imposed by high water tables, slowly permeable soils and shallow in-situ soils (7). Figure 15-3 illustrates the basic design features. In effect the constructed mound raises the application bed above the natural soil surface so that the larger base area of the mound serves as the design infiltration surface for the natural soils. Most regulatory agencies have specific design requirements regarding size, slopes, and construction materials. The standard percolation test is usually accepted for design of the soil mound since infiltration into and out of the mound are controlling parameters. Evaluation of subsequent percolate/groundwater flow in the natural soils will require the procedures discussed in previous sections of this chapter.

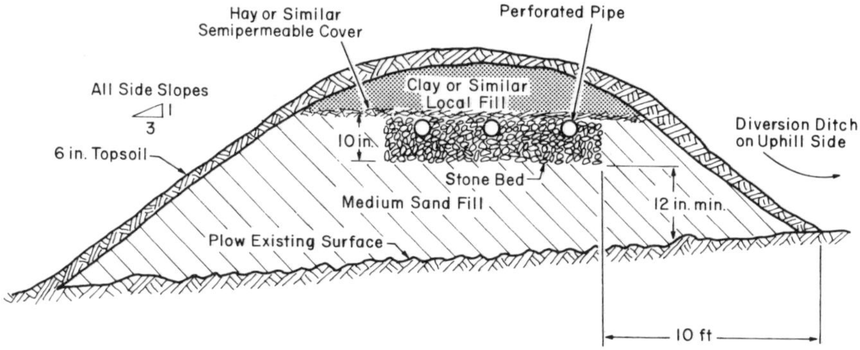

Figure 15-3 Basic design features for soil mound systems.

The basic design approach for a soil mound is a two step operation. Percolation tests are run in the natural soils on the site at the depth of concern. Table 15-3 can then be used to calculate the base area of the mound. Then, based on the type of soil used to construct the mound, the area of the application bed in the mound is determined. Table 15-4 relates the most commonly used fill materials for mound construction to the design rates used for determining the bed area. These values should be confirmed with additional percolation tests after construction is completed and consolidation of the fill has occurred.

Table 15.3: Infiltration Rates for Determining Base Area of Mound (8)

Natural on Site Soil	Percolation Rate (min/in)	Infiltration Rate Q_g (gpd/ft^2)
Sand, sandy loam	0-30	1.2
Loam, silt loams	31-45	0.75
Silt loams, silty clay loams	46-60	0.5
Clay loams, clay	61-120	0.25

Table 15.4: Mound Fill Materials and Infiltration Rates (8)

Material	Characteristics (% by weight)	Infiltration Rate Q_g (gpd/ft^2)
Medium sand	>25%, 0.25-0.2 mm <30-35%, 0.05-0.25 mm <5-10%, 0.002-0.05 mm	1.2
Sandy loam	5-15% clay	0.6
Sand/sandy loam	88-93% sand	1.2

The results of the conventional percolation test are expressed as the number of minutes required for the water level to drop one inch in the test hole after equilibrium has been reached. Table 15-5 can be used to relate these values to the soil permeability descriptors used elsewhere in this text.

Table 15.5: Percolation Rate Related to Other Soil Characteristics

Percolation Rate (minutes/inch)	SCS Descriptor	Permeability Range (inches/hour)
<1	Very rapid	>20
1-10	Rapid	2-20
11-60	Slow-moderate	0.2-2.0
>60	Slow	0.06-0.2

Pressure distribution is essential for all mound systems and is recommended for all other large scale onsite disposal systems.

This will insure uniform application over the entire design area and avoid the sequential failures that have occurred with gravity pipe networks.

References 7 and 8 contain complete details for the design of the pipe networks. These usually consist of a solid pipe manifold connected to a number of evenly spaced perforated laterals.

Rehabilitation of Onsite Systems

Failure is defined as the inability of these systems to move the design quantity of water at the design rate. If failure occurs soon after system start-up it may be due to poor design, poor construction or unanticipated groundwater conditions, or some combination of the three. In many cases, groundwater problems may be resolved by surface regrading to eliminate excess surface water infiltration in the area and/or a relief drain upgradient of the system.

Failures occurring after several years of successful operation may be due to a gradual and unanticipated increase in flow or to biological clogging at the infiltration surfaces. A procedure developed at the University of Wisconsin (9) uses hydrogen peroxide, a very strong oxidizing agent to destroy the organic deposits and restore infiltration capacity.

Lysimeter work at the University of New Hampshire (10) successfully rejuvenated sandy and loamy sand soils which had failed due to the build-up of an organic mat. A 30% solution of hydrogen peroxide and water was successful in all cases. A weaker solution at 7.5% H_2O_2 was also successful for sandy soils. Loading rates used were 0.25 lb H_2O_2/ft^2 of surface, for sands and at least 0.50 lb H_2O_2/ft^2 for silty soils.

Table 15.6: Types and Sources of Data Needed for Design of Small Land Treatment Systems

Data	Principal Source
Wastewater characteristics	Local authorities
Soil type and permeability	SCS soil survey
Temperature (mean monthly and growing season)	SCS, NOAA, local airports
Precipitation (mean and maximum monthly)	SCS, NOAA, local airports
Evaporation and ET (mean monthly)	SCS, NOAA, Ag. extension
Land use	SCS, aerial photography from various sources
Zoning	Local angencies
Agricultural practices	County agent, SCS, Ag extension
Groundwater (depth and quality)	USGS, state agency, local driller's logs
Discharge requirements	State or EPA

SMALL SCALE LAND TREATMENT

Any of the three basic land treatment processes (SR, RI or OF) are suitable for industries and small communities. Table 15-6 lists the type and sources of data required for these small land treatment systems.

Small municipal systems may offer greater flexibility with respect to system ownership and management. For example, contractual agreements (see Chapter 13) with local farmers can be developed to take and use partially treated wastewater. The staffing requirements for the system will depend on the type of system and on the operational arrangements. Figure 14-1 in Chapter 14 illustrates staff requirements for typical municipally owned and operated systems. Table 15-7 presents municipal staff requirements at several small land treatment systems (11).

Table 15.7: Staff Requirements at Small Systems (12)

Location Daily Flow (gpd)	Site Type	Site Control	. . Town/Staff (man days/yr). . . Land Component	Total System
Ravenna, MI 72,000	Open fields	City	7	75
Santa Ana, TX 75,000	Pasture	Farmer owns, city operates equipment	46	100
Wayland, MI	Hay, corn	City owns, farmer harvests	68	172
Winters, TX	Hay	Farmer owned	0	52

Site Identification

A simplified screening procedure can be used to determine if there are parcels of suitable land for the three concepts within a reasonable distance. This procedure can be a desk top analysis using available soils and topographic maps. At the same time, a survey should be started to identify local farmers or land owners who may be willing to participate in the land treatment project. Criteria for this initial screening are presented below. These preliminary land area requirements are very conservative and include allowances for preapplication treatment, storage and unused land. These equations should only be used for initial site screening, not for actual system design.

Slow Rate Systems: The total area required will depend on the number of operating months per year. Most systems will operate between 6 and 12 months per year. The two equations below can be used for those conditions or interpolated for intermediate values (12): Figure 6-2 can be used to determine the operating period.

$$6\text{ months/yr: } A = (2.73 \times 10^{-4})Q \qquad (15\text{-}7)$$

$$12\text{ months/yr: } A = (1.94 \times 10^{-4})Q \qquad (15\text{-}8)$$

where A = total land area required, acres

Q = average daily wastewater flow in community or industry, gpd

Desirable site characteristics for SR are:

Grade: <20% cultivated land

<40% forest or pasture land

Permeability: 0.2 to 0.6 in/h

Soils: Clay loam to sandy loams (GM, SM-d, Mh, Oh, MH)

Depth to groundwater 2 to 3 ft

Parcels of land with these characteristics should be identified and marked on the map. For the typical small community it will not be economical to consider land beyond two to three miles in distance from the town or sites where the pumping head will exceed about 100 ft to reach the site (13). Special circumstances such as no cost for the land or AWT water quality for a surface discharging alternative can justify a greater range for the screening procedure. It is not absolutely necessary that all of the land be in one contiguous parcel, especially if a number of farmers have agreed to participate in the project.

Overland Flow Systems: The equations for estimating total land requirements are (12):

$$6 \text{ months/yr:} \quad A = (1.68 \times 10^{-4})(Q) \qquad (15\text{-}9)$$

$$12 \text{ months/yr:} \quad A = (9.07 \times 10^{-5})(Q) \qquad (15\text{-}10)$$

See Equation 15-8 for definition of terms, Figure 6-2 can be used to estimate the annual operating period. Interpolate for intermediate operating time.

Desirable OF site characteristics are:

Grade: Finished slopes 2-8% (can be constructed on flat terrain).

Permeability: 0.2 in/h or less

Soils: Clay and clay loams (SM-o, Sc, Cl, Ol, CN, OH)

Depth to groundwater - not critical

The "rule of thumb" on cost effective screening distance and elevation for OF are (12): 2-3 miles, 150 ft pumping head.

As in the previous case special site conditions may justify an extension of these values. All parcels of land having these characteristics should be marked on the map. It is not absolutely necessary, but is desirable for the OF system to be on one contiguous parcel since ownership and management is likely to be by the town, and system control and security will be easier with a single site.

Rapid Infiltration Systems: RI systems can typically operate on a year-round basis, so that only one equation is given for estimating total land area required:

$$A = (5.92 \times 10^{-5})\,Q \qquad (15\text{-}11)$$

See Equation 15-8 for definition of terms. As in the other cases this is the total area for the system including an allowance for preapplication treatment, etc.

Desirable site characteristics:

Grade $< 10\%$

Permeability > 0.6 in/h

Soils: sand and sandy loams (Gw, GP, SW, SP)

Depth to groundwater 15 ft

(Non drinking water aquifer)

The suggested limits for distance and elevation difference are: 4-7 miles, 200 ft pumping head.

All parcels of land having these characteristics should be located on the map. It is recommended that the RI system be located on one contiguous parcel of land to reduce costs and allow more efficient operation by the town or industry.

Final Site Selection and Investigation

It is unlikely that the area surrounding a small town or industry will contain a sufficient amount of suitable land for all three of the treatment concepts. One or more is likely to be eliminated in the very early stages of the screening process. A field reconnaissance is suggested in the final stages of the screening process to visually observe and verify the site characteristics identified during the map survey. A simplified ranking procedure can be used in the final selection process. Sites with the lowest land cost, lowest elevation difference and closest proximity to the town will generally rank the highest. A field investigation should then be conducted for each of the potential sites identified in the map surveys.

The first step in the site investigation procedure should be to visit the potential site with a local SCS representative. A few, shallow, hand-auger borings to identify the soil profile should be conducted to confirm the SCS data and check for impermeable layers or shallow groundwater. Infiltration tests are usually only needed for RI sites. A few backhoe pits to 10 ft or more are also recommended for RI sites, but drill holes are usually deferred until preliminary design.

If crops will be grown, a site visit with the county agent or local agricultural or forestry advisor is recommended. The purpose of this site visit is to obtain advice on the type of crops to use and on crop management practices.

Facility Design

Since limited field investigations are conducted, very conservative design criteria are adopted. If the system design requires new facilities for preapplication treatment and storage then a combined pond system is recommended and Chapter 8 should be used for its design.

Hydraulic Loading Rates: It is assumed for this procedure that the LDP for designs are:

- SR systems - Hydraulic capacity of the soil or groundwater nitrogen.
- RI systems - Hydraulic capacity of the soil, since site selection has eliminated sensitive aquifers from consideration.
- OF systems - Hydraulic loadings that will consistently produce secondary, or better, effluent quality.

If some other factor is the LDP then the detailed procedures described in earlier chapters should be used.

SR systems: The design hydraulic capacity is based on the most limiting SCS permeability classification of the soils at the selected site. Figure 15-4 can be used to determine the weekly hydraulic loading for small SR systems. The annual loading is obtained by multiplying this value by the number of operating weeks per year.

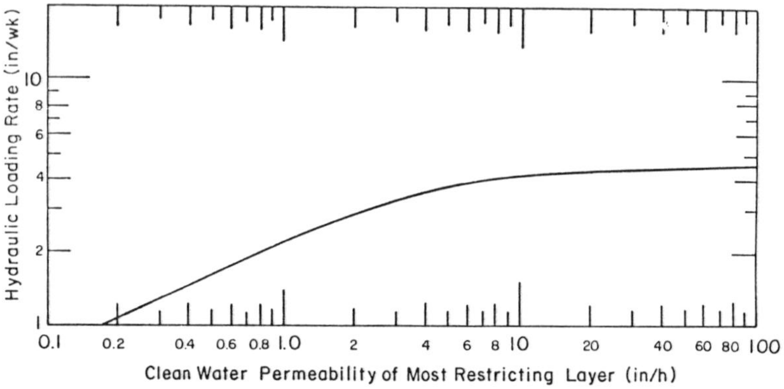

Figure 15-4 Hydraulic loading rate (in/wk) during application season for small scale SR systems, After (12).

The annual hydraulic loading based on nitrogen limits is given by:

$$L_{wn} = 5.2\left[\frac{(2.3)(P_r - ET) + U}{(C_n - 10)}\right] \qquad (15\text{-}12)$$

where L_{wn} = annual hydraulic loading, limited by nitrogen, in/yr

P_r = annual precipitation, in/yr

ET = annual evapotranspiration, in/yr

U = annual crop uptake, lb/acre·yr

C_n = nitrogen concentration in wastewater, mg/L

Table 15-8 can be used to estimate crop uptake (U) for use in Equation 15-12, or a more precise value selected from Chapter 5. Both the crop uptake values in Table 15-8 and the allowance for nitrogen losses in Equation 15-12 are more conservative than the criteria presented for large scale systems in earlier chapters.

Table 15.8: Nitrogen Uptake Rates for Small Land Treatment Systems Design (12)

Crop	Uptake (lb/acre·yr)
Forage crops	
Alfalfa	300
Bromegrass	130
Coastal Bermudagrass	400
Kentucky bluegrass	200
Quackgrass	240
Reed canarygrass	340
Ryegrass	200
Sweet clover	180
Tall fescue	160
Field crops	
Barley	70
Corn	180
Cotton	80
Sorghum	90
Potatoes	230
Soybeans	110
Wheat	60

The hydraulic loading calculated with Equation 15-12 should be compared to the graphical determination from Figure 15-4 and the most conservative of the two used to calculate the actual treatment area required:

$$A = \frac{Q}{(L_w)(3630)} \tag{15-13}$$

where A = SR treatment area, acres

Q = annual wastewater flow, ft^3/yr

L_w = limiting hydraulic loading, in/yr

Additional land is required for preapplication treatment, storage, access roads, and sometimes buffer zones. Allowances for these factors were included in the preliminary screening calculations but specific values must be determined during final design and system layout.

Chapter 9 contains details on wastewater distribution methods. In small communities, it is prudent to choose a distribution method that is used locally or that will result in a system that requires only part-time operational attention. If a locally used distribution method is selected, any specialized equipment and necessary expertise will be more readily available.

Traveling guns require relatively high amounts of labor and are more adaptable to systems where several, odd-shaped fields are irrigated each season, so they are usually owned and operated by a local farmer. Both solid set and center pivot irrigation systems can be adapted to either municipally owned or farmer owned small irrigation systems. Center pivots will generally not be applicable for very small SR systems (below 40 acres).

Typical, small SR systems operate for 8 hours per day for one day per week on a particular plot on a seven day rotation.

Overland Flow: The hydraulic loading rates for small OF systems are essentially the same as defined in Chapter 11. To simplify calculations and area determinations the following criteria are suggested:

	Hydraulic Loading in./wk	Slope Length ft	Operations h/d	Operations d/wk
Screened Wastewater	4	120-150	8-12	5-7
Primary Effluent	5-6	100-120	8-12	5-7
Pond Effluent (when algae is not a concern)	5-7	150	8-18	5-7
Secondary Effluent (activated sludge, trickling filters)	12-16	100-120	8-12	5-7

Storage and operating periods for OF systems can be determined with the procedures in Chapter 11 and 8. Crop selection, distribution methods, runoff collection, etc. are also described in earlier chapters.

Rapid Infiltration: A small RI system need not be designed for intensive wastewater applications at maximum RI rates, which could involve the need for recovery of renovated water and a relatively high level of operation and management. Instead, the design can be simplified to meet the objectives of wastewater treatment and still maintain ease of operation. Figure 15-5 can be used to determine the appropriate hydraulic loading rate depending on the limiting soil permeability and level of preapplication treatment intended. Other design details for RI basin systems can be found in Chapters 9 and 12. A typical operational schedule allows 2 days of flooding followed by 10-18 days of drying with the longer times needed in the winter months. A convenient program uses 2 days of flooding followed by 12 days of drying to complete a 14 day cycle. This would require a minimum of seven cells or basins in the system for continuous operation. Small basins, 0.5 to 2 acres in area are easier to construct and manage for small systems and this will influence the number of cells in the system.

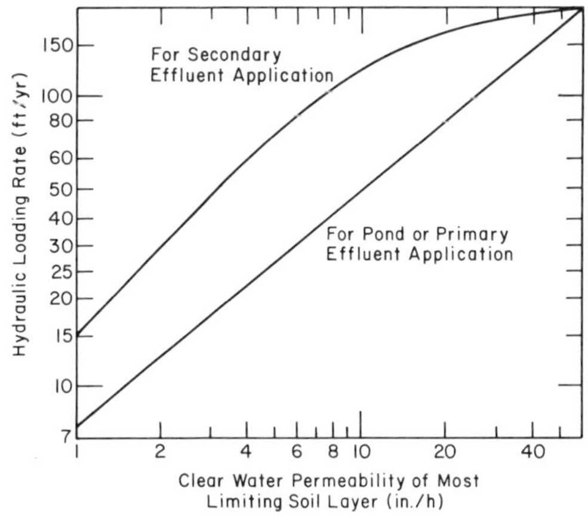

Figure 15-5 Typical annual hydraulic loading rates for small RI systems, After (12).

Seepage ponds have been used successfully in many small communities and are similar to RI in that relatively high hydraulic loading rates are used and treatment occurs as wastewater percolates through the soil. The primary difference is that seepage ponds are loaded continuously, whereas RI systems use a loading cycle that includes both application and drying periods, resulting in improved treatment and maximum long-term infiltration rates. Since the infiltration surface in seepage ponds seldom has an opportunity to dry out the infiltration rate will be retarded. The long-term acceptance rates listed in Table 15-1 should be used for seepage pond design not the values from Figure 15-5. The values in Table 15-1 should be conservative for most cases since they are developed for a 1 ft head in the pond, and greater depths of water should result in higher infiltration rates. Table 15-1 is only valid if biological clogging is the cause of infiltration retardation. Infiltration tests should be run following construction of both RI and seepage basins to verify design assumptions. Inadvertant compaction of the bottom soils during construction can easily cause system failure.

INNOVATIVE CONCEPTS

Innovative technology as discussed in this chapter refers to new concepts or variations as well as proven concepts that have not seen widespread use. This section is only concerned with those innovative concepts that depend on the land and the soil/plant ecosystem as a major component in the system.

Wastewater Systems

A combination of the basic land treatment process (SR, OF, RI) to achieve a particular water quality standard would be considered an innovative approach, some possible combinations were discussed in Chapter 11.

There are a number of possibilities related to the overland flow concept depending on the rate of flow, depth of water and detention time in the system. Grass filtration has been used in

England and Australia since the late 19th century to polish effluent prior to final disposal. A typical operation would apply secondary effluents at relatively high rates to a grass-covered, gently sloping field to obtain further removal of BOD and SS prior to final discharge. As a first approximation the area required might be less than half that required for overland flow designed in accordance with Chapter 11. A further increase in the depth of water and the detention time in a continuous flow system will convert the vegetation to aquatic species and result in a wetland. However, many of the treatment responses will be the same, so the range of possibilities from overland flow to a wetland can be considered as the same group rather than separate and distinct concepts.

Wetland Systems: A significant amount of research and study was directed at the use of wastewater in wetlands and marshes in the 1970's and early '80's (14). Natural wetlands are often associated with, and considered part of, rivers, lakes, estuaries and similar natural water bodies. Since wastewater discharge to these water bodies is typically regulated, discharge to the wetland will have the same constraints. As a result it is not often possible to take full advantage of the treatment potential in a natural marsh since discharge standards must be achieved before wastewater is introduced to the marsh. There are exceptions, and natural wetlands have been successfully developed when the marsh is isolated or when discharge from the marsh to the main water body can be regulated. Typical area requirements for systems of this type are 150 to 450 acres of wetland per million gallons per day (mgd) design flow. Smaller acreages are possible when the main objective is environmental enhancement or restoration.

A major constraint on the use of a natural wetland is the lack of control over the flow regime in the system. Meandering channels may contain most of the flow so the surface area available is not effectively used. This has been partially overcome in some systems with the use of gated irrigation pipe for more uniform distribution at the head of the system.

Constructed Wetlands: The construction of a wetland where one did not previously exist can eliminate the requirement that discharge standards must be achieved prior to the wetland. A constructed wetland is part of the treatment process so discharge standards should apply to the final system effluent. Construction of the unit is similar to OF procedures described in Chapter 11. The bottom can be natural clay or rendered relatively impermeable by compaction or with liners. Soil or other media is then needed to support the aquatic vegetation. Appropriate inlet and outlet structures complete the system.

The vegetation serves several purposes:

- Roots and stems act as support medium for biological growths.
- Leaf canopy shades the liquid surface in summer preventing algae growth.
- Cattails, reeds, and rushes transmit oxygen from the leaves to the roots resulting in aerobic microsites in an otherwise anaerobic environment.

Removal of nitrogen and phosphorus and other elements by the plants is relatively minor so harvest and removal of the plants is not necessary except for special situations. An example might be stringent phosphorus controls. If the plants are cut or allowed to die back in place there could be relatively high P concentration in the spring discharges. Cattails (Typha) have rapidly dominated on most systems regardless of the initial species present. Work in Ontario (15) indicates that a full canopy will develop within a few months if cattail roots are initially planted on about 3 ft centers.

A significant developmental effort has been directed at the

use of constructed wetlands for wastewater treatment in a variety of locations, with different types of wastewater and hydraulic loadings. Some of the most successful results have been obtained by the Ministry of Environment in Ontario, Canada (15). Pilot scale work over several years led to construction of a full scale system in Port Perry, Ontario in 1983. Their pilot scale work revealed several similarities with overland flow experience:

- If high algal content wastewater is applied to the wetland, the algae will not be effectively removed. (Research at the University of California, Davis and at Gustine, CA confirmed this finding.)
- If high solids content wastewater is applied there will be excess deposition and odor problems at the head of the system.
- Winter time operation is possible as long as the soil/ water interface does not freeze.

The most successful performance was with effluent from a partial mix aerated cell applied to a long narrow wetland cell. Wetland effluent was relatively consistent winter and summer:

$$BOD_5 < 15 \text{ mg/L}$$
$$TSS < 10 \text{ mg/L}$$
$$\text{Total N} < 15 \text{ mg/L}$$

Phosphorus removal was limited and chemical additions are necessary to meet stringent P effluent standards. Based on the successful pilot scale work, recommended criteria for a full scale system are:

1. 5 day detention time, partial mix aerated cell (see Chapter 8 for criteria), tapered aeration to permit some settling in final portion of cell.
2. Long narrow wetland channels (L:W = 12:1) with gravel or natural soil for support media, inlet and outlet

structures to maintain water depth at 8 in. in summer and 12-24 in. in winter. Design detention time in summer is 5 days. Winter detention time is dependent on ice thickness but is approximately 10 d.

Based on these criteria a 0.5 mgd system would require about 11.5 acres of wetland surface (5 bays, 100 ft wide by 1,000 ft long). By contrast, a 0.5 mgd OF system designed for secondary effluent might require about 16 acres of treatment surface.

The system is operated at a greater depth in the winter to allow ice formation which in turn acts as an insulation barrier allowing flow to continue under the ice. Apparently the attached growth and debris on the bottom and the long narrow configuration are very effective in particulate removal. The aerated cell will convert much of the BOD to particulate matter even in the winter allowing a high overall removal on a year-round basis.

REFERENCES

1. Healy, K.A., Laak, R., Site Evaluation and Design of Seepage Fields, Journal ASCE EED, 100 (EE5): 1133, (October 1974).
2. Finnemore, E.J., Hantzsche, N.N., Groundwater Mounding Due to Onsite Sewage Disposal, ASCE Journal IDE, 109, (2):199, (June 1983).
3. Fielding, M.B., Groundwater Mounding Under Leaching Beds, Proceedings, Third National Symposium on Industrial and Small Community Sewage Treatment, ASAE, Pub. 1-82, p. 215, (1982).
4. Mott, T.O., et al., Flow Calculation for Household Effluent Disposal in Elevated Sand Mounds, Journal Envir. Qual., 10, (3):311, (1981).
5. Allen, D.H., Hydraulic Mounding of Groundwater Under Axisymmetric Recharge, Research Report 24, University New Hampshire, Water Resources Research Center, Durham, NH., (January 1980).

6. Hantush, M.S., Growth and Decay of Groundwater Mounds in Response to Uniform Percolation, Water Resources Research, 3(1): 227, (1967).

7. U.S.EPA, Design Manual-Onsite Wastewater Treatment and Disposal Systems, EPA 625/1-30-012, U.S.EPA CERI, Cincinnati, OH., (October 1980).

8. Otis, R.J., Pressure Distribution Design for Septic Tank Systems, Journal ASCE EED, 108 (EE1): 123, (1982).

9. Harkin, J.N., Jawson, M.D., Clogging and Unclogging of Septic System Seepage Beds, Proceedings Second Illinois Symposium on Private Sewage Disposal Systems, Illinois Dept. Public Health, Springfield, IL., (1977).

10. Bishop, P.L., Logsdon, H.S., Rejuvenation of Failed Soil Absorption Systems, Journal ASCE EED, 107 (EE1): 47, (1981).

11. U.S.EPA, Process Design Manual-Land Treatment of Municipal Wastewater, EPA 625/1-81-013, U.S.EPA CERI, Cincinnati,OH., (October 1981).

12. Reed, S.C., et al., Wastewater Irrigation for Small Communities in:Proceedings ASCE I&DD, Orlando, FL., July 20-23, p. 501 (1983).

13. U.S.EPA, Generic Facilities Plan for a Small Community: Stabilization Pond, Land Treatment on Trickling Filter, Draft Report FRD 26, U.S.EPA FRD, Washington, D.C., (June 1982).

14. Reed, S.C., et al., Aquaculture Systems for Wastewater Treatment- An Engineering Assessment, EPA 430/9-80-007, U.S.EPA, OWPO, Washington, DC, MCD-68, (June 1980).

15. Black, S.A., et al., Sewage Effluent Treatment in an Artificial Wetland, Presented at WPCF Conference, Detroit, MI, (October 4-9, 1981).

16

Land Application of Sludge

Sludges and other residuals of a land treatment system can be effectively managed by applying these residual materials to the land on a portion of the site. A land treatment system can also be specifically designed for just sludge applications from a municipality or an industry. This chapter only considers those methods that are designed for treatment; landfills and other disposal methods are covered elsewhere (1)(2)(3). There are four major categories of sludge land application systems:

- Agricultural utilization: Sludge is used as a source of fertilizer nutrients and/or as a soil amendment.
- Forest Utilization: Sludge is used to enhance forest productivity.
- Site Reclamation: Sludge is used to reclaim disturbed land, such as strip mined area.
- Soil Treatment: Sludges are incorporated in upper soil layer for treatment by soil organisms. Most common for industrial wastes such as petroleum sludges and toxic and hazardous materials.

The LDP for design of all of these systems are the sludge constituents. For example, the annual application on an agricultural operation may be determined by nitrogen or phosphorus considerations and the useful life of the site limited by cadmium. Since the liquid fraction of sludges is typically low, the hydraulic capacity of the site soils is seldom the LDP for design. An exception might be toxic and hazardous waste sites where very permeable soils would permit too rapid percolate movement.

SITE AND PROCESS EVALUATION

Data will usually be available on the quantity of sludge to be expected or it can be estimated from similar systems elsewhere. A preliminary estimate of the land area required for screening purposes can be determined for municipal sludges with Table 16-1. Estimates of soil treatment area for industrial sludges should be based on the critical LDP with the criteria in Chapter 3.

Site selection follows the same general approach described in Chapter 6.

Table 16.1: Estimated Land Area for Municipal Sludge Applications

Option	Application Period	Reported Range (dry ton/acre)	Typical Rate (dry ton/acre)
Agricultural	Annual	1-30	5
Forest	One application or at 3-5 yr interval	4-100	20
Site reclamation	One application	3-200	50

Slope limitations and recommended set-back distances are summarized in Tables 16-2 and 16-3. Detailed guidance for site selection and evaluation are given in reference (1).

Table 16.2: Recommended Slopes for Sludge Sites (1)

Slope (%)	Comment
0-3	Ideal
3-6	Acceptable for surface application or injection
6-12	Inject liquid sludge
12-15	Immediate incorporation of all sludges. Effective runoff control

Table 16.3: Setback Distances for Sludge Sites

Distance (ft)	Criteria
50-300	Injection, and only near: single dwellings, ponds and lakes, 10 yr high water mark for streams, roads. No surface applications.
300-1,500	Injection or surface application near: all the above, plus springs, and water supply wells; injection only near high density residential developments.
>1,500	Injection or surface application at all the above.

AGRICULTURAL UTILIZATION

The design approach is based on the utilization of sludge as a supplement or replacement for commercial fertilizers. As a result the annual application is based on either the N or P needs of the crop in a particular soil. In addition, both the annual and cumulative sludge loadings must be consistent with regulatory limits on pathogens and metals (Table 3-10 summarizes current EPA limits on cumulative metal loadings for agricultural land). A design approach based on the nitrogen needs of the crop should then impose no greater impact on the groundwater than conventional farming operations in the area with application of commercial fertilizers. As a result groundwater monitoring is not typically required for agricultural systems.

Most states require that soils at sludge application sites be maintained at a soil pH of 6.5 or above. Many states have limits on the cumulative amounts of Pb, Zn, Cu, Ni, and Cd related to the CEC of the soil. Some of these limits differ from the EPA guidance in Table 3-10 so specific values must be obtained in the planning stage for each project. Information on fertilizer recommendations for a particular crop in a specific location can be obtained from the Agricultural Experiment Station in each state, or from local extension personnel. A preliminary estimate of crop nutrient needs can be determined with procedures in Chapter 5. The design is then based on meeting either the N or P needs. Optimum yields and crop production may then require supplemental fertilization for the other nutrients (N,P,K).

Nitrogen Limits

Since much of the sludge nitrogen is in the organic form it is not all immediately available to the plants. A portion will "mineralize" each year and become available, as described in Chapter 3. These contributions must be included in the mass balance for determining the annual application rate. Table 3-13 contains suggested mineralization rates for municipal sludges. Example 3-5 demonstrates the procedure for animal manures and it would be similar for surface applied sludges.

Metal Limitations

Sludges with significant metal concentrations might have an annual application limited by Cd, in this case the N applied would be less than the crop needs. Table 16-4 summarizes the recommended annual Cd limits (1). The annual, or cumulative sludge application based on metal limits is given by:

$$S_m = \frac{M_L}{(C_m)(2{,}000)} \quad (16\text{-}1)$$

where S_m = sludge application rate, for the time interval selected, dry Ton/acre

M_L = metal limitation of concern, an annual basis for Cd, (Table 16-4) or lifetime basis for Cd or other metals (Table 3-11), lb/acre

C_m = percent metal content in the sludge, as a decimal (eg: .005% = .00005) sludge with 50 ppm Cd = .005% Cd.

Table 16.4: Recommended Annual Cadmium Limits (1)

Crop Type	Annual Cadmium* (lb/acre)
Tobacco, leafy vegetables, root crops	0.45
Other food chain crops	1.78 (until 6/30/84) 1.25 (7/1/84-12/31/86) 0.5 (after 1/1/87)
Animal feed only	None

*In all cases, soil pH should be maintained at 6.5 or higher.

Phosphorus Limitations

The design calculation when crop uptake of P is specified as the limiting parameter takes a similar form:

$$S_p = \frac{U_p}{C_p\ (2{,}000)} \qquad (16\text{-}2)$$

where S_p = annual sludge application rate based on crop uptake of P, dry Ton/acre

U_p = annual crop uptake of P, lb/acre (see Table 5-7 for typical values)

C_p = percent available phosphorus in sludge, as a decimal. Assume only 50% of total P in the sludge is available.

The C_p for a sludge with 20 ppm P would be:

$$C_p = (0.50)(0.00002) = 0.00001$$

Sludge Loading Determination

The calculation procedure for sludge loading on a nitrogen basis is a three step procedure:

1. Determine the plant available nitrogen (N_p) in the sludge during the application year:

$$N_p = (2,000)\left[NO_3 + K_v(NH_4) + f_1(ON)\right] \qquad (16\text{-}3)$$

where N_p = plant available nitrogen in sludge during application year, lb/dry Ton of sludge

NO_3 = percent nitrate in sludge, as a decimal

K_v = volatilization factor: 0.5 for surface applied liquid sludge, 1.0 for incorporated liquid sludge and dewatered digested sludge applied in any manner

NH_4 = percent ammonia nitrogen in sludge, as a decimal

f_1 = mineralization factor for first year as a decimal. See Table 3-13. Example : f_1 for digested sludge = 0.2

ON = percent organic nitrogen in sludge, as a decimal

2. Determine the plant available nitrogen (N_{PR}) from mineralization of the residual sludge in subsequent years:

$$N_{PR1} = 2,000 \sum f_2(ON)_2 + f_3(ON)_3 + \ldots\ldots \qquad (16\text{-}4)$$

where N_{PR1} = percent plant available nitrogen from mineralization of the first years sludge in subsequent years, as a decimal

f = mineralization rate (Table 3-13) as a decimal, subscripts refer to the year of concern

ON = percent organic nitrogen remaining in the sludge in a particular year. Subscripts refer to the year of concern, as a decimal

A system with continuous annual applications will have to solve Equation 16-4 for each of the subsequent years, ie: N_{PR2}, N_{PR3}, N_{PR4}, etc. A tabular form is recommended for summation of the plant available nitrogen from all sources for each year. The calculations will converge on a relatively constant value after 5 to 6 years if the sludge composition remains the same.

3. The annual sludge loading (S_{Ny}) based on nitrogen is determined with:

$$S_{Ny} = \frac{U_N}{(\sum N_p + N_{PR1....})} \qquad (16\text{-}5)$$

where S_{Ny} = annual sludge loading in year (y) of concern, dry Ton/acre

U_N = crop uptake of N (see Table 5-7) lb/acre·yr

N_P, N_{PR1} = from Equations 16-3 and 16-4

Land Area Determination

The land area calculation is a five step process:

1. Determine S_N or S_p loading rates depending on state requirements.

2. Determine S_M based on annual cadmium limits.

3. The LDP for design is then the more stringent of step 1 or 2 above.

4. Determine the land area required with:

$$A = \frac{Q_s}{S_{LDP}} \qquad (16\text{-}6)$$

where A = land area required, acres

Q_s = total annual sludge production, dry Ton/yr

S_{LDP} = the limiting sludge loading from step 3 above, dry Ton/acre·yr

5. Use Equation 16-1 and the values in Table 3-10 to determine the useful life of the site.

Example 16-1: Determine Land Area for Application of Digested Municipal Sludge

Conditions:

a. Sludge production: 22 dry T/d

b. Sludge characteristics: Pb = 500 ppm, Cd = 50 ppm, Ni = 100 ppm, Cu = 500 ppm, Zn = 2000 ppm, Total N = 2.5%, NH_4 = 1.0%, NO_3 = 0.

c. Sludge will be incorporated, so K_v = 1, CEC of 10 meq/100g for site soils, Corn is intended crop U_N = 160 lb/acre·yr (Table 5-7).

d. State allows design based on N fertilization rates. Mineralization rates (Table 3-13) for digested sludge:

Year	f
1	0.20
2	0.10
3	0.05
4-10	0.03

Solution:

1. Organic nitrogen (ON) = Total N - NH_4

$$= 2.5 - 1 = 1.5\% = 0.015$$

   a. Available nitrogen first year:

$$N_P = (2{,}000)\left[0 + 1(0.01) + 0.20(0.015)\right]$$
$$= 26 \text{ lb/Ton of sludge}$$

   b. First year's sludge, organic N remaining in second year:

$$ON_2 = (0.015) - (0.20)(0.015)$$
$$= 0.012$$

   Amount of first year's sludge mineralized in second year:

$$N_{PR2} = (f_2)(ON_2) = (2{,}000)(0.1)(0.012)$$
$$= 2.4 \text{ lb/Ton of sludge}$$

   c. First year's sludge, organic N remaining in third year:

$$ON_3 = (0.012) - (0.1)(0.012) = 0.0108$$

   Mineralization of first year's sludge in third year:

$$N_{PR3} = (2{,}000)(f_3)(ON_3) = (0.05)(0.0108)(2{,}000)$$
$$= 1.08 \text{ lb/Ton of sludge}$$

   d. In the fourth year:

$$ON_4 = (0.0108) - (0.05)(0.0108) = 0.0103$$

$$N_{PR4} = (2{,}000)(f_4)(ON_4) = (2{,}000)(0.03)(0.0103)$$
$$= 0.62 \text{ lb/Ton}$$

   e. The N_{PR} for years 5 and beyond converges on a value of about 0.6 lb/Ton

2. Repeat the calculations for sludge applied in years 2 through 5 and tabulate results.

   a. For all applications:

Year After Application	N_{PR} lb/Ton
2	2.4
3	1.1
4	0.6
5	0.6

b.

Year	Total Available N, lb/dry T	Annual Sludge Loading S_N (Eq. 16-5), Dry T/acre·yr
1	26	6.15
2	26 + 2.4 = 28.4	5.63
3	26 + 2.4 + 1.1 = 29.5	5.42
4	26 + 2.4 + 1.1 + 0.6 = 30.1	5.32
5 and beyond	26 + 2.4 + 1.1 + .6 = 30.1	5.32

3. Calculate the annual sludge loading based on Cd:

$$S_M = \frac{1.79}{(0.00005)(2,000)} = 17.9 \text{ Ton/acre}$$

4. For present, nitrogen controls design. If the future cadmium limits in Table 16-4 are retained, Cd would control the annual applications after 1987.

5. Use the "steady-state" S_N to determine land area:

$$A - \frac{(22 \text{ Ton/d})(365 \text{ d/yr})}{(5.32 \text{ Ton/acre·yr})} = 1,509 \text{ acres}$$

6. Determine design life of site:

a. From Table 3-10 for CEC of 10 meq/100 g:

Metal	Cumulative Loading, lb/acre
Pb	1000
Zn	500
Cu	250
Ni	250
Cd	8.9

Design loading = 5.32 Ton/acre·yr

b. Typical computation:

Annual Pb = (0.005)(2,000)(5.32) = 5.32 lb/acre·yr

$$\text{Useful life} = \frac{1000}{5.32} = 188 \text{ yr}$$

c. Summary:

Metal	Useful life, yr
Pb	188
Zn	23.5
Cu	47
Ni	235
Cd	17

7. Therefore, the cumulative Cd loading would limit use of the site to 17 yr with the present criteria.

Monitoring and Application Scheduling

The design example above is based on criteria and regulatory guidance available in late 1983. The values contained in Table 3-10, 3-13 and 16-4 may change significantly so the appropriate regulatory authorities should be consulted prior to any design. The calculation procedure demonstrated will still be valid however. Reference (1) should be used for design of transportation and application procedures. Since the design is based on limiting both N and Cd, monitoring for these parameters should not be necessary. Routine agricultural type soil testing every 2 to 4 years for plant available P and K and to determine lime requirements for pH maintenance at 6.5 is all that should be necessary.

The schedule for sludge applications will depend on the type of crop and on the climate for the area. Sludge is not usually applied when the ground is frozen to reduce risk of runoff losses. Sludge can be applied to the fields for row crops prior to planting and after harvest. Sludge application to forage grasses is possible in all months of the year when the ground is not frozen.

FOREST UTILIZATION

Many aspects of system design for forested sites are similar to the previous case so criteria in Tables 16-2 and 16-3 and the related discussion are applicable. Site options include applications in existing forests or developing a new plantation. In the former case, the sludge will be typically sprayed from specially equipped trucks due to the access difficulties for other techniques. These high pressure sprays may then impose constraints on public access during and after the operation.

Seedlings of some tree species show poor survival when planted directly in freshly applied sludge. It may be necessary to let the sludge age for 6 months or more, to allow for salt leaching and ammonia volatilization.

Seedlings have low nitrogen uptake rates. An intensive program of weed control is necessary since the weeds grow faster than the seedlings, and compete for nutrients, space, light, etc. Use of herbicides and cultivation between tree rows is usually required for the first 3 to 4 years.

Young forest plantations (trees over 2 years old) are more tolerant to sludge applications and weed control is less of a problem. However, sludge spraying may be limited to dormant periods to avoid heavy sludge deposits on the foliage.

The plant nitrogen uptake will be highest in an established forest as compared to the two previous cases. The N uptake will diminish for "old" trees (30 to 60 yr old). Application is possible under the leaf canopy so foliage problems will not occur. The major difficulty with established forests is access. The maximum range of truck mounted spray guns is about 120 ft. Therefore to insure uniform coverage a road grid on about 250 ft centers is necessary or applications limited to the 120 ft strip on either side of existing roads and fire breaks.

Forest Application Scheduling

In many cases it is more typical to apply a large single quantity of sludge every 3 to 5 yr rather than smaller annual applications due to the costs and complexity of transport and access for distribution. As described in the agricultural case, a small annual application will have no greater affect on the groundwater than convention agricultural operations and should be of no greater concern. A large application every 3 to 5 years may result in some of the nitrogen being available for movement to the groundwater via percolation. The nitrogen of concern would be the portion of mineralized organic N (ON) that is not taken up by the plants in the first year. This residual could be nitrified in the soil and move down to the groundwater with the net precipitation falling on the site. An estimate of this quantity can be obtained with the calculations demonstrated in Example 16-1 combined with the procedures in Chapter 10.

The basic design procedures for forested sites is the same as described in the previous case. The plant available nitrogen is determined with Equation 16-3. In this case since application is almost always surface applied liquid sludge, the K_v is equal to 0.5. The plant uptake values for Equation 16-5 can be found in Table 5-10.

Sludge Loading for Forest Sites

There is some variation in the design approach for forested sites. Because trees are not a food chain crop, the annual cadmium limits in Table 16-4 do not apply. In the typical forested project the annual loading will usually be based on nitrogen. There are insufficient data on cumulative metal loadings with respect to toxicity of forest plants. However, it is prudent to adopt some tentative limits for control. In most existing forests the topography and soils probably exclude future conversion to food chain crop land. Therefore, the requirements for maintenance of pH 6.5 and a cumulative Cd loading should not be critical. The following values have recently been suggested by the U.S. EPA for forested sites (1):

Metal	Maximum Cumulative Loading lb/acre
Pb	1000
Zn	500
Cu	250
Ni	250

These are the same as the values in the center column of Table 3-10.

Example 16-2: Determine Land Area for Application of Digested Municipal Sludge in an Established Forest

Conditions: Same as Example 16-1 except: $K_v = 0.5$, U_N = 300 lb/acre·yr (Table 5-10), Use annual applications.

Solution:

1. Nitrogen available in application year:

$$N_p = (2{,}000)\left[(0) + (0.5)(0.01) + 0.20 \quad (0.015)\right]$$

$$\doteq 16 \text{ lb/Ton dry sludge}$$

2. Summary of total available N including mineralized fractions:

Year	Total Available N	S_N
1	16	18.8
2	16 + 2.4 = 18.4	16.3
3	16 + 2.4 + 1.1 = 19.5	15.4
4	16 + 2.4 + 1.1 + .6 = 20.1	14.9
5 and beyond	16 + 2.4 + 1.1 + .6 = 20.1	14.9

3. Land Area $= \dfrac{(22 \text{ t/d})(365 \text{ d/yr})}{(14.0 \text{ T/acre·yr})} = 539$ acres

4. Design life at 14.9 Ton/yr:

   a. Typical calculation:

$$\text{Annual Pb} = (0.0005)(2{,}000)(14.9) = 14.9 \text{ lb/acre·yr}$$

$$\text{Useful life} = \frac{1000}{14.9} = 67 \text{ yr}$$

b. Similarly:

Metal	Useful Life, yr
Pb	67
Zn	8.4
Cu	16.7
Ni	83.9

5. In this case the useful life is limited to 8.4 yr because of the suggested zinc limitations. The cumulative sludge applied to this forested site would be 125 Ton/acre as compared to 90 Ton/acre for the agricultural case in Example 16-1.

If application is limited to the 120 ft strip on either side of existing roads and accessible fire breaks, then about 18.5 mi of such roads would be required.

The application scheduling in forests will depend on the growth stage for the tree and climate for the area. Frozen ground conditions should be avoided. Applications should not be made to young plantations during the growing season to avoid foliage damage. Reference (1) provides detail on transport and application equipment.

SITE RECLAMATION

Extensive areas of disturbed land exist throughout the United States as a result of mining operations. Also fairly widespread are areas where dredge spoils or fly ash have been deposited, and construction areas (e.g., roadway cuts, borrow pits).

Most disturbed lands are difficult to revegetate. These sites generally provide a harsh environment for seed germination and subsequent plant growth. Major soil problems may include a lack of nutrients and organic matter, low pH, low water-holding capacity, low rates of water infiltration and permeability, poor physical properties, and the presence of toxic levels of trace metals. To correct these conditions, large applications of lime and fertilizer may be required, and organic soil amendments and/or mulches also may be necessary. Sludge provides a low cost beneficial substitute for some of these commercial products.

A major distinction between this case and the previous two is that sludge is typically required in large amounts the first year to reestablish fertility and vegetation. As a result the design is based on a one-time application to a particular area. It is therefore necessary that there be a sufficient area of disturbed land available each year of the design life of the project. When mining operations are involved the state and federal regulations on restoration apply in addition to any environmental rules.

The general site considerations discussed previously apply to this case also. Crop selection is more unique since revegetation is often the major goal.

If the aim of the reclamation effort is to establish a vegetative cover sufficient to prevent erosion, a perennial grass and legume mixture is a good crop selection. It is important to select species that are not only compatible, but also grow well when sludge is used as the fertilizer. The rationale for the selection of grass and legume seeding mixtures is that the grass species will germinate quickly and provide a complete protective cover during the first year, allowing time for the legume species to become established and develop into the final vegetative cover. The grasses will also take up a large amount of the nitrogen, preventing it from leaching into the groundwater. Since legume species can fix nitrogen from the atmosphere, additional sludge nitrogen applications are often unnecessary.

Plant species to be used should be selected because of their ability to grow under droughty conditions, and their tolerance for either acid or alkaline soil material. Salt tolerance is also desirable.

If a site is to be reforested, it is still generally desirable to seed it with a mixture of grasses and legumes. The initial grass and legume cover helps to protect the site from erosion and surface runoff, and to take up the nutrients supplied by the sludge. Planting slow growing tree species is generally not recommended because they generally do not compete well with the initial herbaceous cover. Fast growing hardwoods such as hybrid poplars are recommended.

Sludge Application Rates

The basic design approach is to determine the maximum cumulative loading based on metals (Table 3-10) and then apply that volume of sludge in a single application. This approach is conservative in that it protects against possible future conversion to food chain crops even though the original purpose is revegetation. A large initial application is necessary to insure a sufficient pool of nutrients and organic matter for establishment of the vegetation. This high initial loading may result in some nitrate percolation beneath the site in the first year. In many situations the aquifers are not potential drinking water sources. If the state agency is concerned with groundwater quality at the project boundaries then the procedures in Chapter 10 should be used. The input parameters for this calculation would be the unused mineralized nitrogen available in the first year and the net precipitation on the site as the percolate volume.

Example 16-3: Determine Land Area for Application of Digested Sludge for Land Reclamation

Conditions: Same as Example 16-1 except: U_N = 250 lb/acre for grass (Table 5-7).

Solution: Cumulative metal loadings from Table 3-10 and example 16-1 apply.

1. From Table 3-11.

   a. Typical calculation:

      Sludge contains 500 ppm Pb = (0.0005)(2000) = 1 lb/Ton of sludge

$$S_{Pb} = \frac{1000 \text{ (lb/acre)}}{(1 \text{ lb/Ton})} = 1000 \text{ ton/acre}$$

b. Similarly:

Metal	Sludge Loading Dry ton/Acre
Pb	1000
Zn	125
Cu	500
Ni	500
Cd	89

2. The limiting sludge loading would be 89 ton based on the Cd requirements. At a sludge production rate of 22 ton/d (Example 16-1), the required area would be:

$$A = \frac{(22)(365)}{89} = 90 \text{ acres}$$

This much disturbed land would be required for each year of operation. Assuming a 12 yr operation which was the time determined in Example 16-1 the total land required is the same for both cases because the same limitations control.

3. For this case the plant available nitrogen in the first year (assuming incorporation of the sludge) would be equal to 26 lb N/Ton dry sludge (same as Example 16-1).

Total N available = (26 lb/Ton)(89 Ton/acre) = 2314 lb/acre

Crop uptake = 250 lb/acre

Temporary N excess = 2314 - 250 = 2064 lb/acre

In theory all of this is mineralized in the first year and some is potentially available for migration with the percolate. Any impact should occur in the first year since sludge is only applied once and the mineralization rates are lower in subsequent years. In the typical case this temporary one year impact should be acceptable and preferrable to the continued environmental degradation if the site remained unrestored.

Restoration Site Monitoring

Monitoring, for a short period, may be more detailed for this concept than the previous two because of the very large single application.

Background Sampling: Composite soil samples should be collected from the site for the determination of pH, liming requirements, CEC, available nutrients and trace metals prior to sludge addition. Water samples from surface streams, lakes, etc., and private household wells in the area should be analyzed for nutrients and trace metals prior to sludge application. Composite sludge samples should be collected and analyzed to provide data for use in designing loading rates.

Application Sampling: As the sludge is delivered, grab samples should be taken and analyzed for moisture content to adjust the delivered amount of sludge to the design rate if there is variation in the sludge moisture content. Composite sludge samples should be collected as the sludge is applied, to document the actual nutrient and trace metal application rate.

Post Application Monitoring: Monitoring of the sludge application site after sludge has been applied can vary from none to extensive, depending on state and local regulations and site-specific conditions. Generally, it is desirable to analyze the soil after 1 year for soil pH changes and heavy metals (if required). In addition, surface and groundwater analysis for nitrogen forms and trace metals may be needed.

SOIL TREATMENT SYSTEMS

This concept is possible for that group of waste constituents that are amenable to degradation by the biological organisms in the soil (see Chapter 3 for detailed discussion). Since many of these sites are permanently dedicated for this purpose and will never be used for food chain crop production the cumulative metal limitations may not apply. The controlling factor on design loading and scheduling is the rate at which the soil system can degrade the material of concern. The concept can be used for high rate applications of municipal sludge but is most commonly used for industrial sludges and slurries and for toxic and hazardous materials.

Use of the concept for hazardous wastes requires consideration of the cumulative life-time limits for waste constituents in addition to the soil system interactions that control the short-term loading rates (4). Site closure is necessary when the lifetime limits are reached. Planning for ultimate closure is a necessary part of the activities to obtain a permit to open and operate the site. The final surface in most cases must be covered with a permanent vegetative cover to prevent erosion. If metals or other substances have reached phytotoxic levels for this final vegetation, mixing by deep plowing or other neutralization will be needed.

The LDP for design of these soil systems can be determined with the procedures and criteria in Chapter 3. References (4)(5) (6) and (7) also contain similar information on a variety of waste materials. Site preparation in most cases includes:

- Removal of surface vegetation.
- Subdivision of the area into operational plots.
- Construction of runoff control dikes around the entire site.
- Grading to promote surface drainage and collection of runoff for treatment/disposal.

Sludges are usually surface applied at the design loading rate, allowed to dry if necessary, and then incorporated in the top 6 to 12 in. of the soil. The treatment zone may extend to a depth of 5 ft depending on the type of soil, type of treatment expected and on the amount of percolation allowed.

Land treatment of oil refinery wastes has been routinely practiced in the U.S. for over 25 yr. About 1/3 of all the refineries in the U.S. have either full scale or pilot scale land treatment systems (7). Oil reduction at these full scale facilities range from 70 to 90%. Annual loadings range from 70 to almost 2,000 barrels of oil per acre but are typically about 400 BBL/acre·yr. A rate of 400 BBL/acre·yr is approximately 5.8 lb of oil per cubic foot of soil or 7.2% oil in the soil within a 6 in. incorporation depth. This annual loading might be applied in monthly increments or two to three times per year depending on site climate and related operational factors. See Chapter 3 for further discussion on oil as the LDP and for a design example.

REFERENCES

1. U.S. EPA, Land Application of Municipal Sludge, EPA 625/1-83-016, U.S. EPA CERI, Cincinnati, OH., (October 1983).
2. U.S. EPA, Process Design Manual-Municipal Sludge Landfills, EPA 625/1-78-010, U.S. EPA CERI, Cincinnati, OH., (October 1978).
3. U.S. EPA, Process Design Manual-Sludge Treatment and Disposal, EPA 625/1-79-011, U.S. EPA CERI, Cincinnati, OH., (Sept. 1979).
4. U.S. EPA, Hazardous Waste Land Treatment, SW-874, U.S. EPA OSW, Washington, D.C., (April 1983).
5. Parr, J.F., et al., Land Treatment of Hazardous Wastes, Noyes Data Corporation, Park Ridge, N.J., (1983).
6. Overcash, M.R., Pal, D., Design of Land Treatment Systems for Industrial Wastes, Ann Arbor Science, Ann Arbor, MI., (1979).
7. Ryan, J.R., et al., Land Treatment Practices in the Petroleum Industry, American Petroleum Institute, Washington, D.C., (1983).

Conversion Factors

U.S. Customary to SI (Metric)

U.S. Customary Unit Name	Abbreviation	Multiplier	SI Symbol	SI Name
acre	acre	0.405	ha	hectare
acre-foot	acre-ft	1,234	m^3	cubic meter
cubic foot	ft^3	28.32	L	liter
		0.0283	m^3	cubic meter
cubic foot per second	ft^3/s	28.32	L/s	liter per second
degrees Fahrenheit	°F	0.555 (°F-32)	°C	degrees Celsius
feet per second	ft/s	0.305	m/s	meters per second
foot (feet)	ft	0.305	m	meter(s)
gallon(s)	gal	3.785	L	liter(s)
gallons per acre per day	gal/acre·d	9.353	L/ha·d	liters per hectare per day
gallons per day	gpd	4.381×10^{-5}	L/s	liters per second
gallons per minute	gpm	0.0631	L/s	liters per second
horsepower	hp	0.746	kW	kilowatt
inch(es)	in.	2.54	cm	centimeter(s)
inches per hour	in./h	2.54	cm/h	centimeters per hour
mile	mi	1.609	km	kilometer
miles per hour	mph	0.45	m/s	meters per second
million gallons	mg	3.785	ML	megaliters (liters $\times 10^6$)
		3,785.0	m^3	cubic meters
million gallons per day	mgd	43.8	L/s	liters per second
		0.044	m^3/s	cubic meters per second
parts per million	ppm	1.0	mg/L	milligrams per liter
pound(s)	lb	0.454	kg	kilogram(s)
		453.6	g	gram(s)
pounds per acre per day	lb/acre·d	1.12	kg/ha·d	kilograms per hectare per day
pounds per square inch	$lb/in.^2$	0.069	kg/cm^2	kilograms per square centimeter
		0.69	N/cm^2	Newtons per square centimeter
square foot	ft^2	0.0929	m^2	square meter
square inch	$in.^2$	6.452	cm^2	square centimeter
square mile	mi^2	2.590	km^2	square kilometer
		259.0	ha	hectare
ton (short)	ton (short)	0.907	Mg (t)	megagram (metric tonne)
tons per acre	tons/acre	2,240	kg/ha	kilograms per hectare
		2.24	Mg/ha	megagrams per hectare

Other Abbreviations:

second	s	hour	h	week	wk
minute	min	day	d	year	yr

Index